AMPHIBIANS

OF **Oregon, Washington**
AND **British Columbia**

A FIELD IDENTIFICATION GUIDE

BY

CHARLOTTE C. CORKRAN
& CHRIS THOMS

LONE
PINE

To Dave and Richard, for 30 years of companionship
while unearthing insights and expressions
of the mountains and valleys of the northwest,
and for cheerfully ignoring tadpoles and salamanders in the kitchen,

and to you who are naturally drawn to look
into quiet pools of water to know Earth's humble secrets.

The Publisher
Lone Pine Publishing

1901 Raymond Avenue S.W., Suite C	202A, 1110 Seymour Street	206, 10426-81 Avenue
Renton, Washington	Vancouver, British Columbia	Edmonton, Alberta
U.S.A. 98055	Canada V6B 3N3	Canada T6E 1X5

Canadian Cataloguing in Publication Data
Corkran, Charlotte C., 1945–
     Amphibians of Oregon, Washington and British Columbia

     Includes bibliographical references and index.
     ISBN 1-55105-073-0

     1. Amphibians—Northwest, Pacific—Identification. I. Thoms,
Chris, 1943– II. Title.
QL653.N95C67 1996   597.6'09795   C96-910400-6

Senior Editor: Nancy Foulds
Project Editor: Roland Lines
Design: Gregory Brown
Layout and production: Gregory Brown, Carol S. Dragich
Maps: Volker Bodegom, Gregory Brown
All photos and illustrations: Charlotte C. Corkran and Chris Thoms
Separations and film: Elite Lithographers Co. Ltd., Edmonton, Alberta, Canada
Printing: Quality Colour Press, Edmonton, Alberta, Canada

Partial funding for this publication was provided by the Northwest Ecological
Research Institute, Portland, Oregon.
The publisher gratefully acknowledges the support of Alberta Community
Development and the Department of Canadian Heritage.

CONTENTS

ACKNOWLEDGMENTS

Much of the field work that went into preparing this book was funded by the following agencies: Region 6 of the U.S. Department of Agriculture, Forest Service; the Oregon and Washington Office of the U.S. Department of the Interior, Bureau of Land Management; the Wildlife Diversity Program of the Oregon Department of Fish and Wildlife; and the Northwest Ecological Research Institute. We sincerely appreciate their support.

While this book is based on our own observations, we also made use of published material for the size ranges and voice descriptions of several species, the elevation ranges, the descriptions of the eggs of stream-breeding and terrestrial amphibians and other data used to distinguish between species, particularly from Altig (1970), Altig and Ireland (1984), Davidson (1995), Green and Campbell (1984), Hayes (1994), Leonard et al. (1993) and Nussbaum et al. (1983).

We are grateful to the following people for providing us with unpublished information: Dave Clayton (salamanders of southwestern Oregon); Charlie Crisafulli (Larch Mountain Salamander); Marc Hayes (several species, particularly the Spotted Frog); Stan Orchard (several species, particularly the Wood Frog and the Striped Chorus Frog); and Linda Dupuis (the Tailed Frog in British Columbia). Range maps were generally based on Green and Campbell (1984), Kirk (1983), Leonard et al. (1993), Marshall (1993), McAllister (1995), Oregon Natural Heritage Program (1991), St. John (1982 through 1986) and our records. Maps were modified to reflect unpublished distribution records generously provided by Linda Dupuis, Laura Friis (including records contributed to the B.C. Wildlife Branch Database by Pierre Friele, John Kelson, Penny Ohanjanian and Pat Milligan), Jim David, Rick Demmer, Bob Storm and Dana Visalli.

The physiographic province and ecoprovince maps are derived from maps made available by the Wildlife Diversity Program of the Oregon Department of Fish and Wildlife, the Washington Department of Fish and Wildlife and the Wildlife Branch of the B.C. Ministry of Environment, Lands and Parks. The diagram of stream orders is from the U.S. Forest Service Region 6 Stream Inventory Handbook.

We are especially grateful to Gene Silovsky, Barb Hill, Erick Campbell, Claire Puchy, Kelly McAllister, Bob Storm, Maurita Smyth, Teresa DeLorenzo and the other associates of the Northwest Ecological Research Institute for their support, encouragement and advice. This book was improved through extensive reviews of several drafts by John Applegarth, Linda Dupuis, Gary Fellers, Dick Forbes, Laura Friis, John Guetterman, Marc Hayes, Dan Holland, Larry Jones, Bill Leonard, Kelly McAllister, Kristiina Ovaska and Bob Storm, as well as suggestions from many others. We appreciate the cooperation of the Scout Island Nature Center at Williams Lake, B.C., the Hood River Ranger District, the help of Laura Friis and others at the B.C. Wildlife Branch and the Nature Conservancy and Natural Heritage Program offices in Oregon and Washington. Thanks also go to Photocraft and to Dave Clayton, Dave Darda, Rick Demmer, Marc Hayes, Larry Jones, Bill Leonard, Kelly McAllister, Robert Penson, Chuck Peterson (and his students, Steve Burton and Jeff Lacey) and Bob Storm for some great days out there in the mud!

Opposite Page
Juvenile Tiger Salamander
Juvenile Wood Frogs (inset)

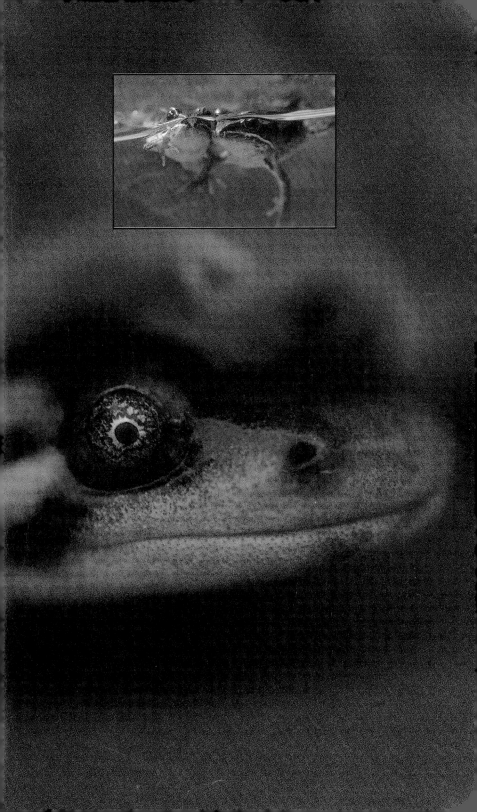

Taxonomy of the Amphibians (Class Amphibia)

ORDER: Caudata (Salamanders)

Aquatic Salamanders

FAMILY: Ambystomatidae (Mole Salamanders)

Northwestern
Salamander
(Ambystoma gracile)

Long-toed Salamander
(Ambystoma
macrodactylum)

Tiger Salamander
(Ambystoma tigrinum)

FAMILY: Salamandridae (Newts)

Roughskin Newt
(Taricha granulosa)

FAMILY: Dicamptodontidae (Giant Salamanders)

Cope's Giant Salamander
(Dicamptodon copei)

Pacific Giant Salamander
(Dicamptodon tenebrosus)

FAMILY: Rhyacotritonidae (Torrent Salamanders)

Cascade
Torrent Salamander
(Rhyacotriton cascadae)

Columbia
Torrent Salamander
(Rhyacotriton kezeri)

Southern
Torrent Salamander
(Rhyacotriton variegatus)

Olympic
Torrent Salamander
(Rhyacotriton olympicus)

ORDER: Anura (Frogs and Toads)

FAMILY: Leiopelmatidae (Bell Toads)

Tailed Frog
(Ascaphus truei)

FAMILY: Bufonidae (True Toads)

Western Toad
(Bufo boreas)

Woodhouse's Toad
(Bufo woodhousii)

FAMILY: Pelobatidae (Spadefoot Toads)

Great Basin Spadefoot
(Spea intermontana)

FAMILY: Hylidae (Treefrogs)

Pacific Treefrog
(Hyla regilla)

Striped Chorus Frog
(Pseudacris triseriata)

Fully Terrestrial Salamanders

FAMILY: Plethodontidae (Lungless Salamanders)

Genus: *Aneides* (Climbing Salamanders)

Clouded Salamander
(*Aneides ferreus*)

Black Salamander
(*Aneides flavipunctatus*)

Genus: *Batrachoseps* (Slender Salamanders)

Oregon
Slender Salamander
(*Batrachoseps wrighti*)

California
Slender Salamander
(*Batrachoseps attenuatus*)

Genus: *Ensatina* (Ensatinas)

Ensatina
(*Ensatina eschscholtzii*)

Genus: *Plethodon* (Woodland Salamanders)

Dunn's Salamander
(*Plethodon dunni*)

Larch Mountain
Salamander
(*Plethodon larselli*)

Van Dyke's Salamander
(*Plethodon vandykei*)

Coeur d'Alene
Salamander
(*Plethodon idahoensis*)

Western Redback
Salamander
(*Plethodon vehiculum*)

Del Norte Salamander
(*Plethodon elongatus*)

Siskiyou Mountains
Salamander
(*Plethodon stormi*)

FAMILY: Ranidae (True Frogs)

Red-legged Frog
(*Rana aurora*)

Cascades Frog
(*Rana cascadae*)

Spotted Frog
(*Rana pretiosa*)

Wood Frog
(*Rana sylvatica*)

Northern Leopard Frog
(*Rana pipiens*)

Foothill Yellow-legged Frog
(*Rana boylii*)

Bullfrog
(*Rana catesbeiana*)

Green Frog
(*Rana clamitans*)

INTRODUCTION

The loss and fragmentation of amphibian habitats in the Pacific Northwest, as well as the alarming decline of some amphibians worldwide, make it imperative for biologists and naturalists to take responsibility for the future of local populations of native species. We hope that an understanding of the life histories, habitat requirements and vulnerabilities of our native amphibians will foster a consideration of their needs during the planning and implementation of projects and developments.

Relatively few people have carried out extensive studies of amphibians in this region, so there are still many unanswered questions about Pacific Northwest amphibians. Besides the fact that new species have just recently been recognized, basic data about the ranges and life histories of several other species have not yet been published. Amateur naturalists and fish biologists have an equal chance with wildlife biologists and research scientists to make 'herpetological history' with their careful observations. Your work in the field can refine and expand both the known ranges and the recorded life histories of amphibians in Oregon, Washington and British Columbia.

ABOUT THIS BOOK

This book describes all the native and introduced amphibian species of Oregon, Washington and British Columbia. It is a practical guide based on our experience, and it is designed to be used in the field by wildlife and fish biologists and professional and amateur naturalists. It complements the more detailed natural history accounts available in other books. Furthermore, it emphasizes the early development stages of amphibians, because on many days, the eggs, hatchlings, larval salamanders and tadpoles are all you can find, but information about identifying them can be hard to obtain. This field guide provides detailed descriptions and photographs of the eggs, hatchlings and larvae (as well as the adults) of each species.

Collecting frogs and salamanders for museum specimens has been a traditional part of surveying amphibian populations. Situations still exist where collecting and preserving specimens is necessary, but species identification is not one of them. Good photographs that show key characteristics can verify your findings and help with your identification questions. This book can help you learn how to identify amphibians in the field with only a hand lens, and how to survey an area without disrupting its essential habitat components.

THE SPECIES DESCRIPTIONS

The descriptions of wide-ranging species are primarily based on individuals we found in northwestern and northcentral Oregon and southwestern and southcentral Washington, because most of our field observations were in those areas. The photographs in this book are of amphibians from many parts of Oregon, Washington and British Columbia (with a few from northern California and Idaho).

The maps show the currently known ranges of the species, to the best of our knowledge, in Oregon, Washington and British Columbia, and they are purposely general to avoid undue prejudice in your consideration of potential species during identifications. If you need more detailed range information, you may find it in some of the sources listed in the references or by contacting the appropriate wildlife agency. The habitat and elevation data are for Oregon, Washington and British Columbia only, and they may not be accurate in other areas.

Two groups of species—the Torrent Salamanders (*Rhyacotriton* spp.) and the Coeur d'Alene and Van Dyke's salamanders (*Plethodon idahoensis* and *P. vandykei*, respectively)—are treated together in our species descriptions and identification keys, because their identification is primarily based on geographic distribution, and there is still some discomfort about the separation of these salamanders into different species. We have provided separate descriptions for the adults in both these groupings.

Egg masses in water (Spotted Frog)

The species accounts and identification keys provide only the most useful characteristics for identifying each species. Some species can be very difficult to distinguish from one another during particular development stages, and these are treated in more detail in the Confusing Species Comparisons. The keys in this book are artificial keys that use only features you can observe in the field without harming the amphibians. (Taxonomic keys, which are organized according to species classification, are based on features that may require killing the animal to be observed.)

Also see the Key to a Sample Quick-reference Box (p. 33).

COMMON AND SCIENTIFIC NAMES

Common names can change from place to place, reflecting local language and knowledge, while scientific names are universal and are used both to identify organisms and to attribute natural relationships within a classification system.

Scientific names, however, may change with new research or new information. When these changes are merely corrections of earlier work, no one argues about them. Changes that split one species into several or that define new genera, however, often cause controversy, because the outcome brings with it a different understanding about natural relationships.

The classification of amphibians in the Pacific Northwest is being refined at this time, and the scientific and common names of the following species have recently become the subject of discussion:

- The small stream salamanders that have yellow undersides are variously called Olympic, Torrent or Seep salamanders, depending on the publication. Everyone agrees on the genus name *Rhyacotriton* for these salamanders, but the single species described in older publications has recently been split into four species.

- There is disagreement over whether or not to reclassify the species commonly called the Pacific Treefrog from *Hyla regilla* to *Pseudacris regilla*. Since members of the genus *Pseudacris* are commonly called Chorus Frogs, the reclassification would also lead to changing the common name from 'Pacific Treefrog' to 'Pacific Chorus Frog.'

- The Spotted Frog (*Rana pretiosa*) has been split into two species (Green et al. 1997).

In naming the amphibian species in this book, we have followed Collins (1990), with the following three exceptions:

- species in the genus *Rhyacotriton* are called Torrent Salamanders
- we have retained the older versions of both the common and scientific names for the Pacific Treefrog (*Hyla regilla*)
- we have retained the older versions of both the common and scientific names for the Striped Chorus Frog (*Pseudacris triseriata*).

It is easy to think of species—and landscapes—as unchanging and ancient. Yet, from an ecological perspective, ours is a youthful and unstable region, freshly emerged from the latest, but probably not the last, advance of continental glaciers. The amphibian fauna of the Pacific Northwest is still evolving; it is intriguing to realize that we may be watching the ongoing radiation of species in the northwest, while new information and new ideas are being published, sifted and argued.

HOW TO USE THIS BOOK

We have tried to write and organize this manual to be easy to use and understand. We have used simple language wherever possible, and any scientific terms that we could not avoid are clearly defined in the glossary. Our definitions and uses of some terms and our ways of measuring features may differ with the uses and methods in other publications, so you should examine the entire manual, including the glossary, before using it in the field.

Although it is fun to look through the photographs first to find the amphibian you are curious about, you will learn more if you go through the following steps as well:

1. Determine the basic type (i.e., salamander or frog/toad) and the life stage of your specimen to decide which key to use. You can refer to the discussions in Amphibian Development (p. 16) and Handling and Measuring Amphibians (p. 19) for help with this determination.

Larch Mountain Salamander

2. Examine the distinguishing features of your specimen carefully to compare it to the descriptions in the key. As you work through the key, you may need to refer to Handling and Measuring Amphibians (p. 19) and the glossary (p. 168).

3. When you think you have identified the species, verify your results by comparing your specimen to the appropriate species account.

4. If you are still uncertain of the species after using the identification keys, but you have narrowed it down to two or three possibilities, the Confusing Species Comparisons (p. 146) may help. These tables describe additional characteristics for species that are very difficult to distinguish from one another during particular development stages.

5. Take photos. If you cannot identify an amphibian, if you think it is an exciting find, or if it is outside the commonly accepted range of the species, take photos of the animal and the habitat in which you found it. Experts at universities, museums and wildlife agencies may help with an identification, and they may be interested in your findings. See Photographing Amphibians (p. 25) and What Data to Collect (p. 163)

Western Toad

You will need to be flexible in using this manual (keeping in mind that taxonomic criteria may not be observable in the field), because amphibians occasionally show up in habitats where they are not supposed to be, or they can have characteristics that are intermediate between those of two defined species. Hybrids rarely occur, however, so when you find an individual with some features of one species and some of another (and you will), go with the majority of features and give greater weight to the first features listed in the keys and confusing species comparisons.

Remember: Careful notes are essential and good photos are invaluable.

WHAT IS AN AMPHIBIAN?

The term 'amphibian'—from the Greek *amphi* ('double' or 'circular') and *bios* (life)—refers to the way most frogs, toads and some salamanders start life in the water in a larval form with gills and later metamorphose or change into a form that can live on land but must return to the water to reproduce. During metamorphosis there is a dramatic change in the animal's appearance, as well as its 'breathing,' feeding, moving and habitat requirements.

Amphibians are highly variable, however, and many species do not fit the original sense of the word 'amphibian.' In the Pacific Northwest, fully terrestrial salamanders lay their eggs on land; they do not go through a free-swimming, aquatic, larval development stage, and their young hatch as fully terrestrial juveniles. Some aquatic salamanders do not metamorphose; they live out their whole life and breed while still in their larval form—a process called neoteny.

The fully terrestrial salamanders in this region do not have lungs; they only 'breathe' through their skin and the lining of the mouth, whereas frogs, toads and metamorphosed aquatic salamanders use all of these mechanisms for respiration.

Amphibians are the only class of vertebrates (animals with backbones) that have no special protective covering on the skin—most fish and all reptiles have scales, birds have feathers, and most mammals have fur. An amphibian's skin is thin, moist and generally smooth, and it is vulnerable to drying, which is why most amphibians live in cool, damp places (although some species have systems that help them retain moisture). Amphibian eggs do not have a shell or membrane around them—a protective layer of jelly maintains moisture around the eggs—and most amphibians lay their eggs in water or in cool, damp places, such as in logs or underground.

Salamanders are predators throughout their lives, while in this region frogs and toads start life as vegetarian tadpoles (although toad tadpoles also scavenge carrion, and spadefoot tadpoles may become cannibals) and do not become predatory until after metamorphosis. Most amphibian species are generalists that will eat anything they can fit into their mouths—the tiniest hatchling salamanders fiercely attack microscopic copepods, while the largest adult amphibians, such as Pacific Giant Salamanders and Bullfrogs, eat mice, fish, ducklings, snakes or other amphibians. Amphibians usually sit and wait for their prey rather than actively chasing it.

In turn, dragonfly nymphs, other carnivorous insects, fish, snakes, raccoons, river otters, herons and other birds eat amphibians at different development stages. Even people eat frog legs. Most amphibians in this region have poison glands that secrete toxic (even to humans), bitter-tasting, slimy or sticky substances. Some garter snakes seem to specialize in eating amphibians, and they have even developed a resistance to the poison of newts.

Many amphibians have an astounding ability to regenerate, or grow back, tails or legs that have been lost. An attacking predator is apt to catch a fleeing salamander just by the tail, in which case the salamander's tail, or part of it, can break off. The salamander can then escape while the predator is distracted by the tail, which may even wiggle for a while. The tail stump soon begins to regrow, and it may be nearly full length again in a few months. The tail of many salamanders is more brightly colored than the body, which may help focus a predator's attention on a part the salamander can live without for a season.

Many amphibians also exhibit defensive displays in response to predation, but we have found that while they usually struggle to get away during normal capture and handling for identification, they generally do not assume defensive postures. We do not

recommend provoking the animals to display the extreme behaviors. On the other hand, one night we saw a Northwestern Salamander put on the full show of defensive behavior in response to a flashlight beam, even though we had not touched the animal.

Although the color pattern of each species of amphibian is usually distinctive, structural features are most reliable for identifying species, because the individuals can vary widely in their actual colors. Furthermore, each individual can slightly change its colors—most frogs and toads look much darker during cold conditions, and tadpoles and aquatic salamander larvae look very pale and dull if they are kept in a container, especially in the dark.

Adult Northwestern Salamander in a defensive posture

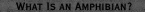

Common Garter Snake

Amphibians and reptiles can look very similar, and they are often grouped together—'herpetology' is the study of amphibians and reptiles—but they are distinctly different classes of animals. Two main features that distinguish reptiles from amphibians are their skin and their eggs.

Western Skink

The skin of a reptile is thick, dry and covered with scales, as opposed to an amphibian's thin, moist, 'naked' skin. Snakes and lizards are scaly all over, while turtles have small scales on their legs and large scales on their shell. Reptiles generally live in warmer, drier places than amphibians because their tough skin helps protect them from drying.

The 'horny toad' is not really a toad at all; that is a common name for the Short-horned Lizard—a reptile.

Most reptiles, including both turtle species in this region, lay their eggs on land, usually in warm, dry places. A hard or leathery shell and an internal membrane keep the eggs from drying out. Hatchlings from reptile eggs are similar to the adults, as are the newborn of reptiles that give birth to live young.

Western Pond Turtle

Western Pond Turtle eggs

Most people know that frogs and salamanders are amphibians, and that snakes and lizards are reptiles. But what about turtles, which can be found on land or in water? Turtles are reptiles.

AMPHIBIAN DEVELOPMENT

Amphibians have different systems of development from egg to adult. Aquatic salamanders lay their eggs in water, and the young hatch as aquatic larvae. In most individuals, the gills and tail fin are later resorbed during metamorphosis into the terrestrial form. Frogs and toads also lay their eggs in water, and they hatch as tadpoles, which appear to be little more than a body with a tail. During metamorphosis, their legs develop and the entire tail is resorbed. The intermediate or in-between stages of metamorphosing salamanders, frogs and toads show characteristics of both the aquatic and terrestrial forms. Fully terrestrial salamanders, on the other hand, lay their eggs on land. The young complete most of their development within the egg and hatch as small versions of the adult form.

The following descriptions of the different amphibian development stages should help you determine which key to use to identify an amphibian. The distinctions between stages used in this book may differ slightly from those in other publications.

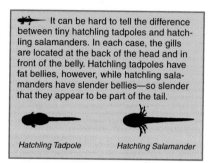

It can be hard to tell the difference between tiny hatchling tadpoles and hatchling salamanders. In each case, the gills are located at the back of the head and in front of the belly. Hatchling tadpoles have fat bellies, however, while hatchling salamanders have slender bellies—so slender that they appear to be part of the tail.

Hatchling Tadpole   Hatchling Salamander

EGGS

Aquatic amphibians lay their eggs in ponds or streams, and each species lays its eggs in a particular type of location. The species also differ in how they lay their eggs: singly, in small clusters or in large jelly masses. The size and color of the eggs or embryos may also be useful in the identification of some species. See the Keys to Eggs (p. 122). When the eggs are near hatching, the embryos themselves can often be identified using the Key to Hatchling Salamanders (p. 128) or the Key to Hatchling Tadpoles (p. 135).

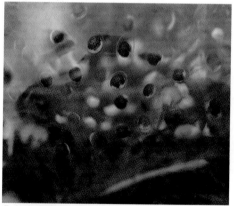

The Keys to Eggs treat only the eggs of amphibians that breed in ponds or slow streams. Amphibians that breed in faster streams and on land conceal their eggs, and the female often tends or guards them. These eggs are rarely found, and we do not recommend searching for them. If you find a concealed nest, quickly replace the covering habitat and record information about the microsite and conditions. There is information on all types of eggs in the species accounts.

Early-stage egg mass (Northwestern Salamander)

SALAMANDERS

Hatchlings: Only pond-breeding salamanders have a hatchling form that is distinct from its larger larval form.

 Stream-breeding salamanders have rarely been documented in their earliest stages of development, and a separate hatchling stage may not occur.

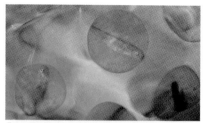
Eggs about to hatch (Northwestern Salamander)

When a pond salamander first hatches it may have no legs, but the front legs begin to develop quickly. For the first week or two, some hatchlings have balancers—a whisker-like appendage on each side of the head—which provide a stable support before the front legs have developed. By our definition, the hatchling stage continues until the nubbins of the hind legs appear. See the Key to Hatchling Salamanders (p. 128). For in-between stages and very small stream larvae, see the Key to Larval Salamanders (p. 129).

Small larva (Northwestern Salamander)

Larvae: Any amphibian with visible gills and four legs is a larval salamander. This stage includes both tiny young

 stream salamanders and huge neotenic forms that have become sexually mature without metamorphosing. See the Key to Larval Salamanders (p. 129). For intermediate stages that still have gill stubs, also see the Key to Metamorphosed and Terrestrial Salamanders (p. 131).

Large larva (Northwestern Salamander)

Juveniles and Adults: This stage includes the metamorphosed forms of pond and stream salamanders that have resorbed

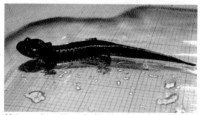
Metamorphosed juvenile (Northwestern Salamander)

their gills and tail fins, as well as terrestrial salamanders, which do not have a larval stage. Sexually immature juveniles are similar to adults, but they have different proportions, and many juvenile terrestrial salamanders have brighter colors that fade as they mature. See the Key to Metamorphosed and Terrestrial Salamanders (p. 131). For intermediate stages that still have gill stubs, also see the Key to Larval Salamanders (p. 129).

Eggs about to hatch (Cascades Frog)

Small tadpole (Cascades Frog)

Large tadpole (Spotted Frog)

*Tadpole with fully developed hind legs
(Northern Leopard Frog)*

Toadlet resorbing tail (Great Basin Spadefoot)

FROGS AND TOADS

Hatchlings: When tadpoles first hatch, they have not yet developed eyes, and they have small gills that are visible at the back of the head (except in Tailed Frogs). Within a few days, however, a covering grows over the gills, and they are no longer visible externally. There may be two nubbins (adhesive glands) under the chin with which the hatchling clings to the egg mass or nearby vegetation. With several species, in warm weather, the gilled hatchling stage may only last a day or two. By our definition, the hatchling stage continues until the gills are covered and the spiracle is developed. See the Key to Hatchling Tadpoles (p. 135). Also see the Key to Tadpoles (p. 138).

Tadpoles: The tadpoles of each frog or toad species have some distinctive features of proportion, shape and color that help with their identification. A tadpole's tail may lengthen considerably just before metamorphosis, however, and some species go through several color stages as tadpoles. Small tadpoles with developed eyes and covered gills can be very difficult to identify; the transparent layer that encloses them is included in the proportions and descriptions in this manual. As the tadpole grows, the hind legs start developing as nubbins at the base of the tail. The front legs develop inside the tadpole's body, and they do not become visible until they pop out fully formed (elbow first and one at a time). The tail is resorbed during the final stage of metamorphosis. You may be puzzled by finding an individual at one of these strange-looking, in-between stages. See the Key to Tadpoles (p. 138). For in-between stages, also see the Key to Frogs and Toads (p. 142).

Juveniles and Adults: The short body and long, folding hind legs of frogs and toads are distinctive, although the existence of a tail remnant can be confusing.  See the Key to Frogs and Toads (p. 142). Juveniles, often called froglets or toadlets, are also included in the Key to

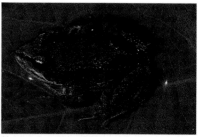
Metamorphosed juvenile (Spotted Frog)

Frogs and Toads, but keep in mind that the juveniles of some species do not show all of the characteristic adult features. The juveniles of many species are discussed separately in the species accounts, and the froglets of Red-legged, Cascade and Spotted frogs are discussed in the Confusing Species Comparisons (p. 151). To identify individuals in intermediate stages (with both a tail and legs), also see the Key to Tadpoles (p. 138).

HANDLING AND MEASURING AMPHIBIANS

You should always be careful not to harm live amphibians while examining them. Here are some basic rules to follow:

- Make sure your hands are wet and free of insect repellent, suntan lotion, or toxins from a previous amphibian.

- Be careful not to let amphibians get overheated or dried out. If your examination takes a long time, moisten your hands periodically, let the animal rest in a plastic bag or covered container, and replace the cold water in the container.

- Clean all field equipment (including plastic bags) after each use.

- Do not detach individual eggs from the mass, nor the mass from supporting vegetation.

- To examine hatchlings, larval salamanders and tadpoles, put them in a clear plastic or glass coin vial or test tube of water or in a plastic bag with a small amount of water.

- Adult salamanders can be examined on your hand, but if your hand is warm the animal will be more agitated. Put the salamander on your work glove (which is usually cold and wet), a leaf or a piece of bark.

- Do not catch or hold a salamander by its tail, because the tail may break off (to regenerate later).

- Frogs and toads are easier to hold if you let the animal extend its hind legs, and then you gently hold the legs together while you support the body.

EGG

jelly layers — egg (ovum)

egg diameter (D)

The methods that follow for identifying and measuring key amphibian features are those that seem to work best for us and for some of the herpetologists with whom we have talked. If you read other amphibian guides, you will notice some differences, since measurement methods are not standardized. Remember that the same person measuring the same live animal several times will get different results. Except in a scientific study, carefully made estimates are just as valuable as precise measurements.

SALAMANDERS

Total length (TL): This is the measurement to use for hatchling salamanders, because it is hard to tell where the body ends and the tail begins. Put the hatchling in water in a plastic bag and hold a ruler underneath the bag. Measure from the end of the snout to the tip of the tail.

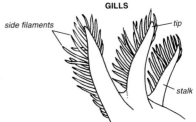

GILLS

side filaments

tip

stalk

Snout-to-vent length (SVL): This measurement is more useful than total length, because part or most of the tail may be missing on many larval and metamorphosed salamanders. Measure from the end of the snout to the front corner of the vent.

This can be very difficult if the salamander is active, or if it coils or retracts into an S-curve. In such cases, one of the following three methods should give at least an estimate of the snout-to-vent length:

- Put the salamander in a plastic bag with a little water and hold the ruler underneath the bag. Use a hand lens if necessary to locate the vent. Terrestrial salamanders can be held in the crease of a wet plastic bag and gently pushed into a straight position.

- For a terrestrial or metamorphosed salamander, let it walk on some surface and then block its progress or even gently tap it on the nose to stop it. Hold the ruler along or just above its back, and measure from the snout to the rear edge of the hind legs, which is approximately above the front of the vent. A flexible plastic ruler can be bent to follow the curve of an uncooperative individual.

- Put the salamander and a short ruler on a flat surface (or in a flat container with a little water for aquatic forms). When the salamander is relatively still and straight, take a photograph from directly above. Make sure the ruler will be able to be read in the photograph. You can then compare the salamander to the ruler on the photograph.

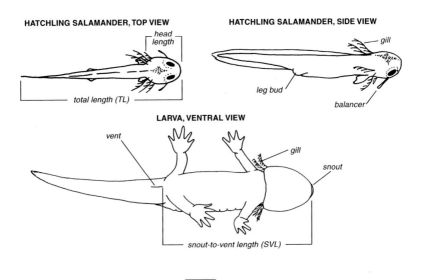

HATCHLING SALAMANDER, TOP VIEW

head length

total length (TL)

HATCHLING SALAMANDER, SIDE VIEW

gill

leg bud

balancer

LARVA, VENTRAL VIEW

vent

gill

snout

snout-to-vent length (SVL)

Leg length: The leg-to-body proportion, which is a diagnostic characteristic of many species, is determined by adpressing the limbs. You adpress the limbs by pressing the hind leg forward and the front leg back (on the same side), while keeping the body from bending. Whether the toes overlap, touch or do not meet is a distinguishing characteristic of several species.

This method must have been developed on dead or sedated animals, because we have never met a person with the manual dexterity or a salamander with a cooperative attitude for this to be physically possible with a live animal. But don't despair, there are two ways to 'visually adpress' the legs:

- Photograph the salamander from directly above when both legs on one side are held straight out to the side (rather than down away from the camera). This is best done in a flat container with water barely covering the animal. The photograph can then be used to estimate the leg-to-body proportions.

- Place the salamander in a shallow container. Watch how far back on the body the front leg reaches as the salamander walks around, and take note of a color marking or scar that marks that point. Do the same for the hind leg on the same side, or try physically adpressing it, and compare that to the front leg's reach.

Counting costal grooves and intercostal folds: These counts are difficult, but they can distinguish between many similar species. Make a count by gently confining the salamander in the crease of a wet plastic bag. Use a hand lens to look through the bag and count the grooves. If you need to visually adpress the limbs, use a method described above. You can take photos through the bag, but take several from different angles to avoid reflections and glare off the bag.

Toe counts and measurements: Use a hand lens to make sure you do not miss seeing the very small inner or outer toes on some species. Toes are always numbered starting from the inside, nearest the body—the 'thumb' or 'big toe' is #1.

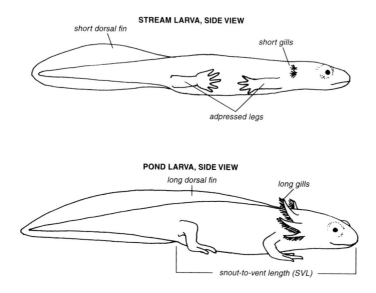

STREAM LARVA, SIDE VIEW

short dorsal fin

short gills

adpressed legs

POND LARVA, SIDE VIEW

long dorsal fin

long gills

snout-to-vent length (SVL)

Nasolabial groove: This feature is present only on some terrestrial salamanders, and it is hard to see, particularly on tiny individuals. With a hand lens and the maximum available light, you can often see a break in the reflection of light across the slight groove if you turn the animal slightly back and forth.

nasolabial groove

[**Gill raker counts:** The number of gill rakers is a diagnostic characteristic for identifying larval salamanders. Gill rakers can only be counted on a specimen that has been killed, however, so it is not a technique for field identification.]

METAMORPHOSED ADULT SALAMANDER, TOP VIEW

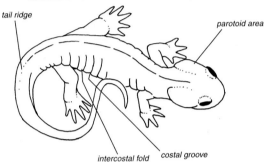

tail ridge

parotoid area

intercostal fold   costal groove

TADPOLES

Snout-to-vent length (SVL): The easiest way to estimate the snout-to-vent length is to view the tadpole from above, using the body outline to determine where the body ends. (In this manual, the term 'body' refers to the head and body of a tadpole, including the transparent layer that surrounds them during early development stages.)

Tail length: This measurement refers to the length of the tail muscle and fin. To avoid confusion about where the tail begins, the measurement should be taken while viewing the tadpole from directly above; measure from where the body ends to the back tip of the tail fin.

Vent orientation: The vent is usually too small to see on a hatchling, but the body outline may reflect the orientation of the vent. Put the hatchling in water in a small coin vial, a test tube or the corner of a new plastic bag. With a hand lens, look up from directly underneath. If the hind end of the body is slightly longer and thicker on the hatchling's left side, it is an indication that the vent may open to the right side; if the hind end appears symmetrical, then the vent may open straight back. You will need to tilt the hatchling back and forth very slightly several times to make sure you are viewing from directly underneath. You may also be able to look from directly above, slightly tilting the hatchling back and forth to look around the dorsal fin.

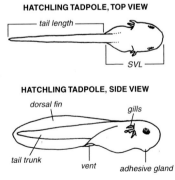

HATCHLING TADPOLE, TOP VIEW

tail length

SVL

HATCHLING TADPOLE, SIDE VIEW

dorsal fin

gills

tail trunk

vent

adhesive gland

Large tadpoles can be corralled in the corner of a plastic bag with a small amount of water. Viewed from underneath with a hand lens, you may be able to see whether the opening of the vent is aimed straight back or to the tadpole's right. An easier way to determine this is to put the tadpole in a small, shallow container of water (a coin vial or test tube is even better) and watch for it to pass material from its intestines. Since tadpoles are usually plant-eating machines, you should not have long to wait.

Tooth rows: It is difficult to count the number of tooth rows on a live tadpole. If you put the tadpole in water in a coin vial, a test tube or the corner of a clean plastic bag, you can use a hand lens to look in its mouth when it opens. At the least, you can see if there are only a few tooth rows or as many as five or six. If necessary, you can hold the tadpole out of water for a few seconds, count rapidly when it gasps, and then quickly return it to the water. Small tadpoles do not have fully developed tooth rows, and they cannot be used to identify most species until the hind legs are well developed.

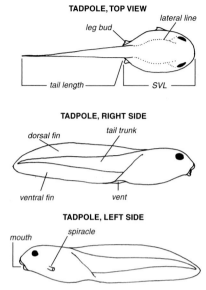

TADPOLE, TOP VIEW

lateral line
leg bud
tail length
SVL

TADPOLE, RIGHT SIDE

dorsal fin
tail trunk
ventral fin
vent

TADPOLE, LEFT SIDE

spiracle
mouth

TADPOLE MOUTH

tooth rows

FROGS AND TOADS

Snout-to-vent length (SVL): As used in this manual, the snout-to-vent length is equivalent to the total length of the head and body. Hold the frog or toad with the ruler aligned along the belly (this may be easiest while holding one hind leg extended), and gently press down on the back. Measure from the tip of the snout to the end of the body (not including the tail on male Tailed Frogs).

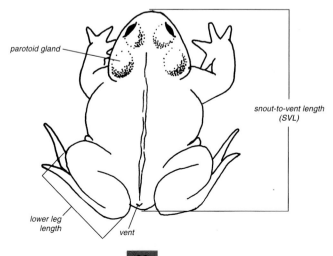

parotoid gland

snout-to-vent length (SVL)

lower leg length

vent

Eye orientation: Determining whether the eyes are oriented more upwards or more out to the sides is difficult.

The eyes are oriented upwards if:

- when looking down from directly above, the eyelids appear to be crescent-shaped, covering a small portion of the eyeball, and you can easily see the pupil and the portion of the iris below the pupil of both eyes at the same time, and

- when looking head-on, the eyeball protrudes as high above the head as the eyelid.

eyes oriented up

The eyes are oriented out to the sides if:

- when looking down from directly above, the eyelids appear to be semicircular or three-quarter-moon-shaped, covering half or more of the eye, and you can not wholly see both eyes at the same time, **and**

- when looking head-on, the eyelid protrudes above the eyeball.

eyes oriented out

Photographs from directly above can be used to compare with the photographs and drawings in this manual.

Leg length: While the hind leg is folded, accurately measure the length of the lower leg (from the knee to the heel). Whether this measurement is more or less than half the snout-to-vent length is sometimes used to distinguish between species.

Determining the leg length by the traditional method (extending one hind leg forward under the body) can harm the animal. If the frog struggles, let go of that leg and then try again.

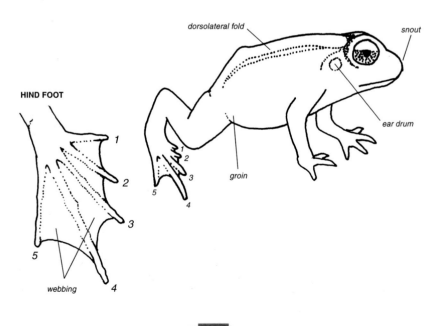

HIND FOOT

dorsolateral fold

snout

ear drum

groin

webbing

PHOTOGRAPHING AMPHIBIANS

When you find an amphibian that you cannot identify in the field, or when you find a species outside its known range, bring back good photographs rather than the animal. Good photos can be examined at length, and they can be taken to an expert for help in identification. They can also be used as vouchers in museum collections.

A camera and color film should be part of your basic equipment for amphibian surveys. A close-up (macro) lens is a good investment (and essential for voucher photos), and a flash is also very helpful. The photos must be clear and close, and they must clearly show all the identifying features of the species. Photos of the habitat are also useful. (For further information, see other works listed in the references.)

The following is a minimum list of the photos you should take:

- **Whole animal from above:** For salamanders, make sure that the animal is relatively straight, the snout is not tucked down, and the legs are out to the sides rather than under the body. Patience is a must. For frogs, it is fine to have the hind legs extended and being held.

- **Whole animal from the side:** Photograph tadpoles from the left side to show the spiracle. Photograph frogs with their legs extended, to show the groin color.

- **Close-up of the top of the head and gills:** Photograph from straight above to show the eye orientation, proportions and overall shape.

- **Close-up of the hind foot:** Photograph salamanders with the foot out to the side, so that the toes are spread. For frogs, it may be necessary to have another person hold the tips of the inner and outer toes and spread the foot to show the extent of webbing.

- **Underside:** This is important for frogs. If the hind legs are being held, make sure your hand does not obscure the underside of the thighs. A frog will often remain still if you hold it upside down and gently stroke its belly for several seconds.

Larval salamanders and tadpoles can be photographed in a wide, shallow plastic tub or a clear plastic bag (with the top rolled down for views from above). Water distorts the image, however, so use barely enough to cover the animal, but allow the gills to float. Take several photos, each from a slightly different angle, to assure that at least one is free from reflections. It may be easier to photograph frogs and toads with another person to hold the animal.

AMPHIBIAN HABITATS

Amphibian species in the Pacific Northwest occur in a wide range of habitats because of different habitat preferences during their various development stages. While fully terrestrial salamanders appear to develop within a limited range of terrestrial habitat elements, in other amphibian species the young start life in an aquatic form and later change into a form that lives on land.

Regardless of the number of development stages, each species requires specific habitat elements that provide enough food and cover, and suitable temperature and moisture levels. These needs change with the different development stages and seasons, and many have several mechanisms that permit them to take advantage of a variety of conditions. An example is a species with both neotenic (larval) and metamorphosed forms of sexually mature adults, which are able to survive in totally different habitats from each other.

Most amphibians are opportunistic and flexible in their habitat use, and we sometimes find them in the most surprising places—in an urban basement, under a pumpkin, in tire ruts, even in parking lots. These are not optimal conditions, however, and habitat alteration is thought to be one of the primary causes of declines in amphibian populations. Disruptions can leave animals in habitat remnants or force them into other poor habitats where egg laying may occur, but the successful development of the offspring to maturity may be prevented by deficiencies in critical habitat elements.

Wetland restoration and habitat improvement projects are typically designed without knowledge of the specific habitat requirements of amphibians. Commonly, even the most well-intentioned projects inadvertently establish conditions that favor the expansion of unwelcome species that displace our native frogs and salamanders.

Recent wetland restoration

Established restoration site

For the purposes of this book, we have divided the amphibian habitats of the Pacific Northwest into nine general habitat types.

MARSH AND WET MEADOW (Palustrine—Wetlands)[1]

Marshes, wet meadows, mountain bogs, shrubby or forested wetlands, swales and wet old-fields; also includes very slow, marshy parts of streams; often a mud or organic muck bottom.

SHALLOW POND (Palustrine—Wetlands)

Temporary, ephemeral or permanent water less than 1 m deep at high water, including ponds, lake edges, vernal pools, pockets of water in wet meadows, stock ponds or tanks and temporarily flooded fields; also includes very slow, shallow parts of streams with little or no flow; usually a mud or organic muck bottom.

DEEP POND (Palustrine—Wetlands, and Open Water)

Permanent water 1–3 m deep (or deeper) at high water, including deep ponds, lake or reservoir edges, beaver ponds, oxbows and stream backwaters; also includes deep parts of very slow streams with little or no flow; usually a mud or organic muck bottom.

---

[1] The names in parentheses are the wetlands and deepwater habitats of the Cowardin classification system (Cowardin 1979) that, in general, correspond to these habitat types.

SMALL STREAM (Riverine—Upper Perennial, and Intermittent)

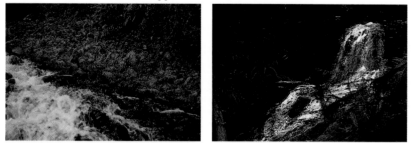

Channel and bank of first or second order[2] (headwater) streams, which are cold, very steep, fast flowing and either perennial or intermittent; also includes springs, seeps and splash zones around waterfalls; a cobble, gravel, boulder or bedrock bottom with no silt.

MEDIUM STREAM (Riverine—Upper Perennial)

Channel and bank of second to fourth order streams, which are cool, moderately steep, moderately fast flowing and perennial; usually a cobble, gravel, boulder or bedrock bottom with little or no silt.

LARGE STREAM (Riverine—Upper Perennial)

Channel and bank of third to fifth order streams, which are warm or cool, less steep, moderately slow flowing and perennial; usually a cobble, gravel, boulder or bedrock bottom with little silt.

---

[2] Stream orders follow the U.S. Forest Service Region 6 classifications (see p. 164).

FOREST AND LOGS

Conifer or hardwood forests with large downed logs.

TALUS

Moist or dry rocky areas with gravel-, cobble- or boulder-sized talus, often at the base of a cliff; also includes fractured rock outcrops.

GRASS AND SHRUB

Dry meadows, brushy areas, shrub/steppe plant communities and dry old-fields.

PHYSIOGRAPHIC PROVINCES OF OREGON

OCR	Oregon Coast Range
KM	Klamath Mountains (Siskiyou Mountains)
WIV	Western Interior Valleys
WSC	West Slope Cascades
ESC	East Slope Cascades
CB	Columbia Basin
BM	Blue Mountains (Ochoco, Blue, and Wallowa mountains)
HLP	High Lava Plains
B&R	Basin and Range
OU	Owyhee Uplands

SPECIES	OCR	KM	WIV	WSC	ESC	CB	HLP	BM	B&R	OU
Northwestern Salamander	■	■	■	■	▲					
Long-toed Salamander	■	▲		■	▲	▲	▲	■	▲	
Tiger Salamander					▲	▲			▲	▲
Roughskin Newt	■	■	■	■	▲	▲	▲			
Cope's Giant Salamander	▲			▲	▲					
Pacific Giant Salamander	■	■	▲	■	▲		▲			
Columbia Torrent Salamander	▲									
Southern Torrent Salamander	▲	▲		▲						
Cascade Torrent Salamander				▲	▲					
Clouded Salamander	■	■	▲	■						
Black Salamander		▲								
Oregon Slender Salamander				▲						
California Slender Salamander		▲								
Ensatina	■	■	■	■	▲					
Dunn's Salamander	■	▲	▲	■	▲					
Larch Mountain Salamander				▲	▲					
Western Redback Salamander	■			■	▲					
Del Norte Salamander		▲								
Siskiyou Mountains Salamander		▲								
Tailed Frog	■	▲		■	▲			▲		
Great Basin Spadefoot					▲	▲	▲	▲	■	▲
Western Toad	▲	▲	▲	▲	▲	▲	▲	▲	▲	▲
Woodhouse's Toad						▲				▲
Pacific Treefrog	■	■	■	■	▲	■	■	■	■	■
Red-legged Frog	■	■	■	■	▲					
Cascades Frog				■	■					
Spotted Frog			?	▲	▲			▲	■	▲
Northern Leopard Frog						?	?			?
Foothill Yellow-legged Frog	▲	■	▲	▲						
Bullfrog	▲	▲	■	▲	▲	■	▲	▲	▲	▲

0   125
kilometres

30

PHYSIOGRAPHIC PROVINCES OF WASHINGTON

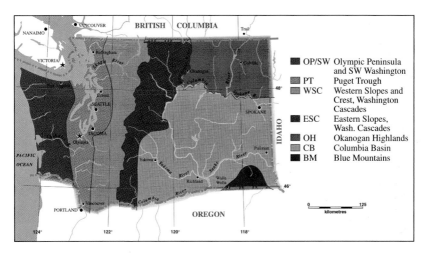

OP/SW	Olympic Peninsula and SW Washington
PT	Puget Trough
WSC	Western Slopes and Crest, Washington Cascades
ESC	Eastern Slopes, Wash. Cascades
OH	Okanogan Highlands
CB	Columbia Basin
BM	Blue Mountains

SPECIES	OP/SW	PT	WSC	ESC	CB	OH	BM
Northwestern Salamander	■	■	■	▲			
Long-toed Salamander	■	■	▲	■	■	■	■
Tiger Salamander				▲	▲	▲	
Roughskin Newt	■	■	■	▲			
Cope's Giant Salamander	■	▲	▲				
Pacific Giant Salamander	▲	▲	■	▲			
Olympic Torrent Salamander	▲						
Columbia Torrent Salamander	▲						
Cascade Torrent Salamander		▲	▲				
Ensatina	■	■	▲				
Dunn's Salamander	▲						
Larch Mountain Salamander			▲	▲			
Van Dyke's Salamander	▲	▲	▲				
Western Redback Salamander	■	■	▲				
Tailed Frog	■	▲	■	▲			▲
Great Basin Spadefoot				▲	▲	▲	▲
Western Toad	▲	▲	▲	▲	▲	▲	▲
Woodhouse's Toad				▲			
Pacific Treefrog	■	■	■	■	■	■	■
Red-legged Frog	■	■	■				
Cascades Frog	■		■	▲			
Spotted Frog		▲		▲	▲	■	■
Northern Leopard Frog					▲	?	
Bullfrog	▲	■	▲	▲	■	▲	▲
Green Frog		▲				▲	

LEGEND

■ = occurrence in suitable habitat in most of the province
▲ = occurrence in a few localities or in one part of the province
? = old records, no recent sightings in the province

ECOPROVINCES OF BRITISH COLUMBIA

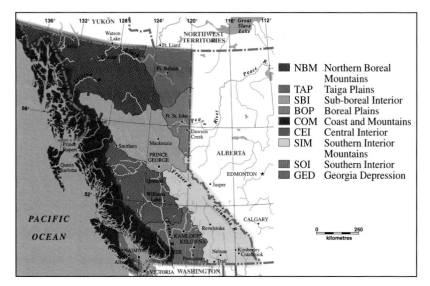

SPECIES	COM	GED	NBM	SBI	CEI	SOI	TAP	BOP	SIM
Northwestern Salamander	■	■							
Long-toed Salamander	■	■	▲	■	■	■		■	■
Tiger Salamander						▲			
Roughskin Newt	■	■							
Pacific Giant Salamander	▲	▲							
Ensatina	▲	■							
Clouded Salamander	▲	■							
Coeur d'Alene Salamander									▲
Western Redback Salamander	▲	■							
Tailed Frog	▲	■							▲
Great Basin Spadefoot					▲	▲			
Western Toad	▲	▲	▲	▲	▲	▲		▲	▲
Pacific Treefrog	▲	■			▲	■			■
Striped Chorus Frog							▲	▲	
Red-legged Frog	▲	■							
Spotted Frog	▲	▲	■	■	■	■		▲	■
Wood Frog	▲	▲	■	■	▲		■	■	▲
Northern Leopard Frog						?			▲
Bullfrog	▲	▲				▲			
Green Frog	▲	▲							

LEGEND

■ = *occurrence in suitable habitat in most of the ecoprovince*
▲ = *occurrence in a few localities or in one part of the ecoprovince*
? = *old records, no recent sightings in the ecoprovince*

KEY TO A SAMPLE QUICK-REFERENCE BOX

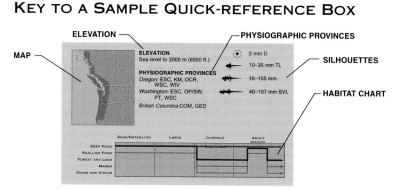

The **MAP** shows the currently known range of the species in Oregon, Washington and British Columbia. It does not show the species's occurrence outside this region.

ELEVATION lists the range of elevations at which the species has been found in Oregon, Washington and British Columbia.

The **HABITAT CHART** shows the different habitat types in which the species may be found during its different development stages. It shows the species' preferred habitats (dark red), as well as habitats in which it is less likely to be found (light red, or blue for neotenic adults).

The **SILHOUETTES** show the different development stages for each species, as well as the approximate size at each stage. For eggs, the measurement is the average diameter (D), not including any jelly layers; for hatchling salamanders it is the minimum to maximum total length (TL); for all other stages it is the minimum to maximum snout-to-vent length (SVL). All these measurements are in millimeters.

PHYSIOGRAPHIC PROVINCES lists each species's occurrence by physiographic provinces in Oregon and Washington and by ecoprovinces in British Columbia. The names and abbreviations are from the Oregon Natural Heritage Data Base, the Washington Department of Fish and Wildlife and the Wildlife Branch of the B.C. Ministry of Environment, Lands and Parks.

OREGON

B&R	Basin and Range	**KM**	Klamath Mountains
BM	Blue Mountains		(Siskiyou Mountains)
	(Ochoco, Blue and Wallowa mtns.)	**OCR**	Oregon Coast Range
CB	Columbia Basin	**OU**	Owyhee Uplands
ESC	East Slope Cascades	**WSC**	West Slope Cascades
HLP	High Lava Plains	**WIV**	Western Interior Valleys

WASHINGTON

BM	Blue Mountains	**OP/SW**	Olympic Peninsula
CB	Columbia Basin		and Southwestern Washington
ESC	Eastern Slopes,	**PT**	Puget Trough
	Washington Cascades	**WSC**	Western Slopes and Crest,
OH	Okanogan Highlands		Washington Cascades

BRITISH COLUMBIA

BOP	Boreal Plains	**SOI**	Southern Interior
CEI	Central Interior	**SIM**	Southern Interior Mountains
COM	Coast and Mountains	**SBI**	Sub-boreal Interior
GED	Georgia Depression	**TAP**	Taiga Plains
NBM	Northern Boreal Mountains		

A Q U A T I C

Southern Torrent Salamander

SALAMANDERS

NORTHWESTERN SALAMANDER

Ambystoma gracile (BAIRD, 1857)

Adult

WHEN *threatened, an adult Northwestern Salamander will lift its back up and tip its snout down (like a bucking horse) to present its poison glands to an attacker. The poison is not very toxic, but the large, white drops that ooze out and run down the salamander's sides are alarming. Both the neotenic and metamorphosed forms of adult Northwestern Salamanders can become stunningly large. Some neotenes and juveniles of this species show little evidence of the poison glands, which can be confusing when trying to identify them. The larvae of this species are the most variable in our region, and we have found ourselves saying, 'If we aren't positively certain, it must be a Northwestern.' • Amby·stoma means 'blunt mouth'; gracile can mean 'slender' or 'simple,' but since this species is hefty, gracile may refer to the simple or plain color pattern of the metamorphosed adult.*

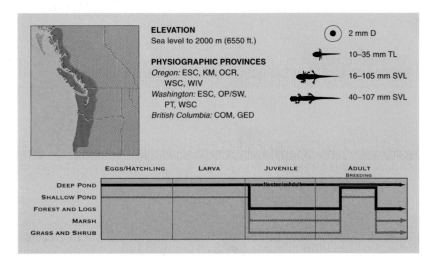

ELEVATION
Sea level to 2000 m (6550 ft.)

PHYSIOGRAPHIC PROVINCES
Oregon: ESC, KM, OCR,
  WSC, WIV
Washington: ESC, OP/SW,
  PT, WSC
British Columbia: COM, GED

2 mm D

10–35 mm TL

16–105 mm SVL

40–107 mm SVL

	EGGS/HATCHLING	LARVA	JUVENILE	ADULT BREEDING
DEEP POND			Neotenic Adult	
SHALLOW POND				
FOREST AND LOGS				
MARSH				
GRASS AND SHRUB				

HABITATS: Northwestern Salamanders live in moist forests or partly wooded areas. They breed in permanent ponds, beaver ponds or stream backwaters in early to mid spring. They lay their eggs in water, and they attach them to a small-diameter stick or rigid stem, 0.5–2 m below the water surface. Hatchlings live in surface sediments or under small debris. Larger and neotenic larvae live under submerged logs or in surface sediments, usually in water deeper than 0.5 m. Metamorphosed juveniles and adults live underground. They can be found aboveground at night during and after rain, and sometimes in soft logs or in bark and wood mounds around the bases of snags.

EGGS are massed in a very firm ball of jelly, which has a smooth or slightly wavy surface because of the additional jelly layer around the entire mass, not just the individual eggs. Egg masses are small orange– to small grapefruit–sized. When first laid, the individual eggs are tan above and cream or pale gold below. Green algae often grow inside the inner jelly layer of each egg, and in time they may color the egg mass.

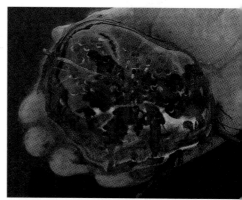

Egg mass

HATCHLINGS have a streamlined outline. The head is small (approximately one-fifth of the hatchling's total length), and the eyes look out to the side. The gills appear somewhat rigid, and they are usually held out from the body at 45° (top view). The few side filaments of the gills are all the same length, and they occur along the full length of the gill stalk. The balancers are light-colored, and they are usually lost before the front legs are fully formed. Hatchlings look long and slender until the hind legs develop, which is usually when a hatchling is at least 25 mm TL. (Also see Confusing Species Comparisons, p. 146.)

Hatchling

Small larva

LARVAE are extremely variable. The head is large, with a broad snout. The gills resemble ostrich plumes because the side filaments are long and willowy, and occur along the entire gill stalk. The poison glands, visible as yellow dots on most large larvae (about 40 mm SVL), are concentrated at the parotoid areas and along the tail ridge within the transparent fin. Larvae are generally olive brown, with large, dark spots on the back and fins. They are most likely to metamorphose during their second summer, but some metamorphose during the first summer. Neotenic Northwestern Salamanders can grow very large. (Also see Confusing Species Comparisons, p. 147.)

Larva

METAMORPHOSED JUVENILES AND ADULTS have a chunky build. They are almost uniformly brown, except that the poison glands are conspicuously lighter than the body. The poison glands are concentrated at the parotoid areas and along the back and tail ridge. The costal grooves are pronounced. There is a confusing form of metamorphosed Northwestern Salamander, however, that does not exhibit the conspicuous glands or costal grooves. These individuals are smaller and darker-colored than the form usually encountered.

Juvenile

Neotenic adult

LONG-TOED SALAMANDER

Ambystoma macrodactylum BAIRD, 1849

Adult

THE *Long-toed Salamander is wide ranging, with several subspecies, and it may be the most versatile amphibian in the Pacific Northwest. It occurs in wet coastal forests, cold mountain meadows and dry steppes. Long-toed Salamanders often breed in newly formed, temporary or recently disturbed pools of water. This is the earliest species to breed each year, which may help the tiny hatchlings avoid predation by the larvae of larger salamanders and insects that are not active until later. Early breeding has its own risks, however, and one year when we visited a breeding pool we found 80 adults and hundreds of eggs that had all died during a sudden, long period of sub-freezing weather.* • Amby·stoma *means 'blunt mouth';* macro·dactylum *means 'long toe,' and it refers to the long fourth toe, which is the diagnostic characteristic of this species.*

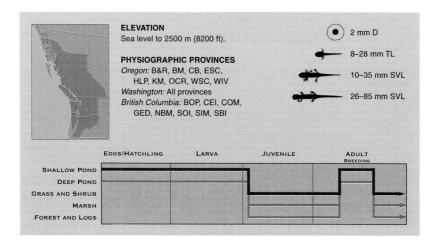

ELEVATION
Sea level to 2500 m (8200 ft).

PHYSIOGRAPHIC PROVINCES
Oregon: B&R, BM, CB, ESC, HLP, KM, OCR, WSC, WIV
Washington: All provinces
British Columbia: BOP, CEI, COM, GED, NBM, SOI, SIM, SBI

2 mm D

8–28 mm TL

10–35 mm SVL

26–85 mm SVL

	EGGS/HATCHLING	LARVA	JUVENILE	ADULT BREEDING
SHALLOW POND				
DEEP POND				
GRASS AND SHRUB				
MARSH				
FOREST AND LOGS				

HABITATS: Long-toed Salamanders live in various habitats, including grasslands, woods and disturbed areas. They breed in seasonal pools, shallow lake edges or very slow streams through wet meadows in winter to early spring. They lay their eggs in water, singly or in clusters, and attach them to a fine stem, a leaf or a pebble (or occasionally on mud) in water less than 0.5 m deep. Hatchlings and larvae live in surface sediments or under rotting leaves, logs or rocks in shallow water. Juveniles can be found under rocks at the edges of ponds in mid summer. Adults stay underground, and they may be found under rocks and logs during the rainy season.

EGGS are large, and each is surrounded by a thick jelly layer (wider than the egg diameter) that makes the eggs appear widely spaced within the cluster. An egg and its jelly layer is usually more than 10 mm in diameter. West of the Cascades, the eggs are laid singly or evenly spaced in a small, soft mass that is the size of a small plum and is often attached to a stem. Within the Cascades and to the east, they are laid

singly or occasionally in a small cluster, often attached to a rock or vegetation, and they may be widely scattered. Individual eggs are black or brown above and white or cream below.

Egg masses

HATCHLINGS have large heads that are approximately one-third of the hatchling's total length and appear much wider than the body. The snout is broad and the eyes look out to the side. The lower gills bend out from the body, often at 90° (top view), while the inner gills bend up above the head. The tip of the gill stalk is a long spike, and the side filaments are shortest near the tip of the stalk and longest near its base. The balancers are long, have a dark tip and are usually lost after one week. Hatchlings develop quickly, and the hind legs usually appear before a hatchling is 25 mm TL. (Also see Confusing Species Comparisons, p. 146.)

Hatchling, with gills curved out

Small larva

LARVAE have a large head with a broad snout. The tip of the gill stalk is long, but it may be lost from some gills. The gill filaments are usually short, and they appear graduated in length, with the longest filaments near the base of the gill stalk. The hind toes are longer than the palm of the hind foot. Larvae are often translucent buff or tan. They may have a fine, dark surface network, but they lack large spots. Larvae develop quickly, and they generally metamorphose during the first summer (except at high elevations) at less than 30 mm SVL. (Also see Confusing Species Comparisons, p. 147.)

METAMORPHOSED JUVENILES AND ADULTS have long hind toes, especially the fourth, and the costal grooves are generally distinct. Metamorphosed salamanders are black or dark gray, often with a pinkish tone on the legs and underside. The sides and underside have white flecks. The dorsal stripe is yellow or green, and it is irregular and broken but has sharp edges.

Larva

Metamorphosing larva

Adult

TIGER SALAMANDER

Ambystoma tigrinum (GREEN, 1825)

Small adult

THE *Tiger Salamander is one of our largest species of salamander, in either the neotenic or metamorphosed form of the adult. The magnificent gills of a larva, often longer than its head, catch one's attention with their shimmer of blue and gold iridescence. The larvae were extensively used for fishing bait, and their release at the end of an unsuccessful day of fishing may explain several small, isolated populations. On the other hand, this is a tough amphibian—it can walk long distances during wet weather, and it can withstand alkaline water that would pickle most animals—and its distribution may be entirely natural. • Amby-stoma means 'blunt mouth'; tigrinum means 'like a tiger,' and it reflects both the bright yellow-and-black striping of some adults and the ferocious predation of this species, which, like most amphibians, will eat any animal it can fit in its mouth.*

ELEVATION
100–1000 m (330–3300 ft).

PHYSIOGRAPHIC PROVINCES
Oregon: B&R, CB, ESC, OU
Washington: CB, ESC, OH
British Columbia: SOI

2 mm D

11–30 mm TL

16–150 mm SVL

55–162 mm SVL

	EGGS/HATCHLING	LARVA	JUVENILE	ADULT BREEDING
DEEP POND			Neotenic Adult	
SHALLOW POND				
GRASS AND SHRUB				
MARSH				

HABITATS: Tiger Salamanders in our region live in grasslands and shrub-steppes. They breed in warm ponds (either permanent or retaining water until mid summer) or shallow lake edges in mid to late spring. They lay their eggs in water less than 1 m deep, usually attached to a branch or stem. Hatchlings and larvae live in aquatic weeds, under logs or in organic sediments in shallow water. Neotenic larvae live in warm ponds and shallow lake edges. Metamorphosed juveniles and adults generally stay underground, but they may be found above ground at night during rainy periods.

EGGS are laid singly, but often so close together that some eggs may touch. They are brown or gray above and cream below, and each is encased in a thin layer of jelly. An egg and its jelly layer is usually less than 10 mm in diameter.

HATCHLINGS have huge gills—the gill stalks are thick and longer than the length of the head—that are swept back along the body (top view). Viewed from the front, the lower pair of gills point down and extend below the body. There are no true balancers, but the lower pair of gills may act in their place. The hind legs usually do not appear until a hatchling is at least 25 mm TL.

Eggs on sedge leaves

Hatchling, with gills swept back

Close-up of gills

Larva

LARVAE have large heads (approximately one-third of the snout-to-vent length). The gill stalks are longer than the length of the head and very wide, and they have short side filaments all the way to their tips. On very large larvae, the toes are wide at the base and taper to points, and the eyes are tiny compared to the huge head. Neotenic adults are common, but only in permanent water.

METAMORPHOSED JUVENILES AND ADULTS have a large, round head and tiny eyes. The toes are distinctly wider at the base and taper to points. The body has a bold mottling of greenish, yellowish, cream or olive tan patches on a brown or black background. The patches are large, regular and often interconnected.

Juvenile

ROUGHSKIN NEWT

Taricha granulosa (SKILTON, 1849)

Adult

NEWTS have the most toxic poison of any amphibians in our region, and people have died from accidentally boiling a newt in the camp coffeepot. Newts rarely release their poison, however, and instead rely on their bright color to warn away predators. Both as larvae and as adults, newts have seemingly endless appetites. They regularly eat the eggs out of the masses of Northwestern Salamanders. Whole groups of eight or more mating newts have been seen clasping each other in a large ball. We know of several ponds with hundreds of newts but no other amphibians (and not even ducks). • Taricha means 'a preserved mummy' (we have no idea why); granul·osa means 'full of small grains,' and it refers to the grainy skin.

ELEVATION
Sea level to 1900 m (6250 ft).

PHYSIOGRAPHIC PROVINCES
Oregon: CB, ESC, HLP, KM, OCR, WSC, WIV
Washington: ESC, OP/SW, PT, WSC
British Columbia: COM, GED

		2 mm D
		8–15 mm TL
		8–38 mm SVL
		12–80 mm SVL

	EGGS/HATCHLING	LARVA	JUVENILE	ADULT BREEDING
DEEP POND				
SHALLOW POND				
MARSH				
FOREST AND LOGS				
GRASS AND SHRUB				

HABITATS: Roughskin Newts live in forested, partially wooded and developed areas. They breed in ponds, lakes or stream backwaters, often where there is abundant aquatic plant growth, including water very stained by rotting vegetation. They lay their eggs in water 0.5–2 m deep in mid spring through summer, and they attach them to fine stems or floating vegetation. Hatchlings and larvae live in vegetation, surface sediments or under debris. Juveniles and adults live in or under soft logs. They forage on the forest floor during damp conditions, even during the day. Some adults may remain in ponds year-round.

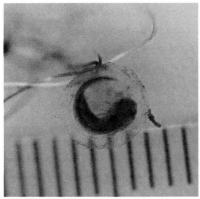

Egg about to hatch

EGGS, which are laid singly, are common, but they are hard to find because they are usually well hidden, possibly under a folded leaf or tucked at the base of radiating stems. The eggs are tan above and cream below, and they are encased in a thin layer of jelly (thinner than the egg diameter).

HATCHLINGS have a small head (approximately one-fifth of the hatchling's total length). The eyes look forward—the front corners of the eyes are closer together than the back corners. The balancers are short. The front legs develop rapidly, and they are very long and skinny. The upper part of the front leg (shoulder to elbow) is longer than the lower part (including the foot), so the feet and elbows are usually visible out to the sides of a hatchling, not hidden beneath its gills (viewed from above). Hatchlings may have one or more rows of yellow dots along their sides. The hind legs usually appear when the larva is less than 15 mm TL. (Also see Confusing Species Comparisons, p. 146.)

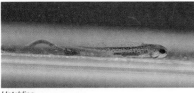

Hatchling

Tiny larva

Tiny larva

LARVAE have a small head with a narrow snout. From the front, a larva looks like it is pouting, because the corners of the mouth turn down. The eyes look slightly forward, and they are pale and often crossed by a dark bar. The gills look ragged, because

Larva

the stalk and filaments are speckled, and the side filaments are of uneven lengths. The tips of the gill stalks often droop down. The front legs are very long and skinny. Larvae are generally translucent tan, with several rows of yellow dots high along the sides. The underside is orange or pink, more distinctly so on larger larvae. Larvae often metamorphose at less than 25 mm SVL, even in the second summer. (Also see Confusing Species Comparisons, p. 147.)

METAMORPHOSED JUVENILES AND ADULTS have a grainy skin surface. There are no costal grooves. The eyes are pale yellow, and are crossed by a distinct, dark bar. There is a strong contrast between the dark brown of the back and sides and the bright orange of the underside. Adults that remain in ponds have a less grainy skin texture, they are paler (their undersides may be gray or cream), and they appear puffy, as if they stayed in a bathtub too long.

Adult

COPE'S GIANT SALAMANDER

Dicamptodon copei NUSSBAUM, 1970

Neotenic adult

COPE'S *Giant Salamander is almost always neotenic—only four metamorphosed adults have ever been documented. There are still many unanswered questions about Cope's Giant Salamanders, and some people remain unconvinced that it is a separate species from Pacific Giant Salamander. Both species sometimes occur in the same streams, but Cope's Giant Salamander usually occurs higher up in the stream system. Most commonly, it is in the same small, cold, fast streams as Tailed Frogs and Torrent Salamanders. If you are examining several individuals that you have caught, put the small ones in a separate bucket or they may suddenly disappear (as lunch for the larger ones!).* • Di·campt·odon *means 'double, bending teeth'; copei honors E.D. Cope, a famous paleontologist in the late 1800s.*

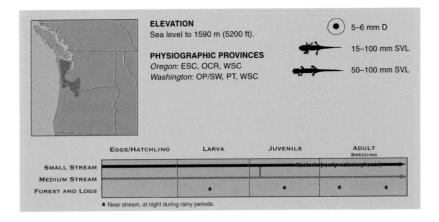

ELEVATION
Sea level to 1590 m (5200 ft).

PHYSIOGRAPHIC PROVINCES
Oregon: ESC, OCR, WSC
Washington: OP/SW, PT, WSC

5–6 mm D

15–100 mm SVL

50–100 mm SVL

	EGGS/HATCHLING	LARVA	JUVENILE	ADULT BREEDING
SMALL STREAM				Neotenic (rarely metamorphoses) →
MEDIUM STREAM				
FOREST AND LOGS		★	★	★ ★

★ Near stream, at night during rainy periods.

HABITATS: Cope's Giant Salamanders live in small, steep-gradient, permanent streams with clear, cold water. The streambed, composed of large gravel to small boulders, with some large logs, has no silt. The eggs may be hidden in or under large logs or rocks in a stream, or under a streambank. They are probably laid in spring and fall. Larvae live under cobbles or small, flat boulders, often at the edge of the main flow. They forage on streambanks and in streamside forests on rainy nights. There are very few records of metamorphosed adults; those individuals that do metamorphose probably live along rocky streambanks.

EGGS are rarely found. They are individually attached in groups to the inside of a log or crevice, and the female protects them. The eggs are not colored.

HATCHLINGS: No distinct hatchling stage has been documented.

Small larva

LARVAE have a long body, and the toes do not touch if the legs are adpressed. The head is distinctively rectangular, flat and long (approximately one-quarter of the snout-to-vent length). It appears to be as narrow at the back of the head near the gills as at the front of the head near the eyes, and it is almost as narrow as the body. The gill filaments are shorter than the gill stalks, so the gills are separate and look like shaving brushes or bundles of larch needles. The short dorsal fin is low (side view). It begins on the tail behind the hind legs and the vent, and the top edge slopes up gradually from midway down the length of the tail. The poison glands may appear as golden, grainy or pitted patches in the skin of the head, back and tail. Larvae are usually brown-gray, with lavender-gray undersides. There is often white around the vent, or a narrow whitish stripe down the middle of the underside. Cope's Giant Salamanders almost always mature as neotenic larvae. (Also see Confusing Species Comparisons, p. 148.)

Large larva

METAMORPHOSED JUVENILES AND ADULTS are rare. The body is long and stocky, and the toes do not touch if the legs are adpressed. Metamorphosed salamanders are brown or gray, with long, tan or metallic patches (similar to a paisley print) that may be interconnected.

Large larva, head

PACIFIC GIANT SALAMANDER

Dicamptodon tenebrosus GOOD, 1989

Neotenic adult

THE *Pacific Giant Salamander is the heaviest salamander in our region—it grows to more than 30 cm long, whether neotenic or metamorphosed (only the Tiger Salamander is as long). Finding yourself nose to nose with the huge head of a metamorphosed adult can be startling, but they rarely bite; instead they generally struggle to return to their hiding places in the rocks. Curved scars or large chunks missing from their tails are evidence of their battles with each other. They reportedly growl if they are harassed. We once cut open a road-killed Pacific Giant Salamander and found four snails and a whole shrew in its stomach. • Di·campt·odon means 'double, bending teeth'; tenebrosus means 'dark' or 'gloomy,' and it may refer to the dark corners beneath the logs and rocks they inhabit.*

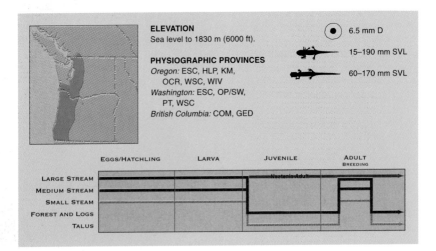

ELEVATION
Sea level to 1830 m (6000 ft).

PHYSIOGRAPHIC PROVINCES
Oregon: ESC, HLP, KM,
    OCR, WSC, WIV
Washington: ESC, OP/SW,
    PT, WSC
British Columbia: COM, GED

6.5 mm D

15–190 mm SVL

60–170 mm SVL

	EGGS/HATCHLING	LARVA	JUVENILE	ADULT BREEDING
LARGE STREAM			Neotenic Adult	
MEDIUM STREAM				
SMALL STEAM				
FOREST AND LOGS				
TALUS				

HABITATS: Pacific Giant Salamanders live in small and mid-sized streams and adjacent moist forests. The streambed, composed of large gravel to small boulders, with lots of large logs, has little silt. The eggs may be hidden in logs or under rocks or streambanks. They are probably laid in spring and fall. Small larvae are often found in small side streams. Larvae live out of the turbulent flow, under cobbles, small (often flat) boulders or logs, or they may be exposed in a pool. Larvae forage on the bank on rainy nights. Metamorphosed adults occur among rocks and logs at the edge of a stream, or in talus that has water flowing under it. During rainy periods, they may be found under forest litter, often at some distance from water.

EGGS are rarely found. They are individually attached in groups to the inside of a log or a crack in a rock, and the female protects them. The eggs are not colored.

HATCHLINGS: No distinct hatchling stage has been documented.

LARVAE have a stout body that is not very long, while the legs are stout and long, so that the toes touch or overlap if the legs are adpressed. The head is massive and wedge-shaped or rounded. It is noticeably wider at the back of the head near the gills than at the front of the head near the eyes, and it is noticeably wider than the body. The gills are small and bushy— the side filaments are longer than the stalks and often hide them. The dorsal fin is tall (side view). It begins on the lower back near the hind legs, and the top edge arches up from near the base of the tail. Larvae are gray-brown. The underside is usually whitish, but it may be gray with white around the vent. There are black markings on the dorsal fin. Pacific Giant Salamanders often mature as neotenic larvae, which may be huge. (Also see Confusing Species Comparisons, p. 148.)

Small larva

Larva

Large larva

METAMORPHOSED JUVENILES AND ADULTS have a wide head that is massive or domed on large individuals. The body is very chunky and not very long, and the toes touch or overlap if the legs are adpressed. Metamorphosed salamanders are brown or gray, with a bold mottling of brassy or coppery, interconnected, long patches (similar to a paisley print).

Metamorphosed adult

TORRENT SALAMANDERS

Rhyacotriton spp.

ALSO CALLED SEEP SALAMANDERS; FORMERLY THE OLYMPIC SALAMANDER

Adult Southern Torrent Salamander

THE *four Torrent Salamander species are generally identified by their geographic location. The observable differences between them are subtle, and the species have been split based on genetic differences that require more than a hand lens. Most field biologists would have a hard time distinguishing between the species if they were all put in the same bucket. Torrent Salamanders soon die if the water is warmed by more than a few degrees.*
• Rhyaco·triton *means 'small stream, god of the sea';* olympicus *refers to the Olympic Mountains;* cascadae *refers to the Cascade Mountains;* variegatus *means 'a variety of colors or forms';* kezeri *honors J. Kezer, emeritus professor at the University of Oregon.*

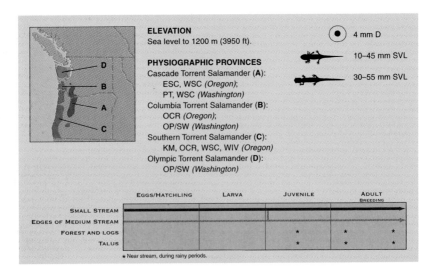

ELEVATION
Sea level to 1200 m (3950 ft).

PHYSIOGRAPHIC PROVINCES
Cascade Torrent Salamander (**A**):
ESC, WSC *(Oregon)*;
PT, WSC *(Washington)*
Columbia Torrent Salamander (**B**):
OCR *(Oregon)*;
OP/SW *(Washington)*
Southern Torrent Salamander (**C**):
KM, OCR, WSC, WIV *(Oregon)*
Olympic Torrent Salamander (**D**):
OP/SW *(Washington)*

● 4 mm D

10–45 mm SVL

30–55 mm SVL

	EGGS/HATCHLING	LARVA	JUVENILE	ADULT BREEDING
SMALL STREAM				
EDGES OF MEDIUM STREAM				
FOREST AND LOGS			★	★ ★
TALUS			★	★ ★

★ Near stream, during rainy periods.

HABITATS: Torrent Salamanders live in very cold, clear springs, seeps, headwater streams and waterfall splash zones, and they may forage in moist forests adjacent to these areas. They lay their eggs in rock crevices in seeps, mostly in the spring. Larvae and adults live in gravel or under small cobbles in silt-free, very shallow water that is flowing or seeping. Adults may also be found under debris on streambanks or in streamside forests and talus during rainy periods.

EGGS have been documented from only a few nests, in which the eggs were laid singly and unattached. The very large, white eggs had several jelly layers.

HATCHLINGS: No distinct hatchling stage has been documented.

LARVAE have a small head, and the eyes are close to the end of the snout. The gills are tiny and hair-like, with few or no visible side filaments. The tail has a small fin. Larvae are translucent honey tan in color, often with 'salt-and-pepper' flecks. The underside is bright yellow or orange-yellow.

Larval Cascade Torrent Salamander

METAMORPHOSED JUVENILES AND ADULTS have huge eyes that are comically perched near the end of the very short snout. Males have a distinct raised ridge on each side of the vent. Metamorphosed salamanders are generally translucent honey tan in color (or they may look greenish tan), often with white or silver flecks. The underside is conspicuously yellow or orange-yellow. Four species of Torrent Sala-mander are now recognized. • **Cascade Torrent Salamander** (*R. cascadae* Good and Wake, 1992) occurs in the Cascade Range. It typically has black flecks on the back, side and belly. • **Columbia Torrent Salamander** (*R. kezeri* Good and Wake, 1992) occurs in the Coast Range north of Little Nestucca and Grand Ronde Valley. It lacks the black flecks, and it has a straight but blurred border between its orange-tan back and orange-yellow underside. • **Southern Torrent Salamander** (*R. variegatus* Stebbins and Lowe, 1951) occurs in the Coast Range south of Little Nestucca and Grand Ronde Valley. It usually has lots of black flecks on the back and underside, especially among individuals from the north end of its range. • **Olympic Torrent Salamander** (*R. olympicus* [Gaige, 1917]) occurs on the Olympic Peninsula. It has black spots on the belly and a sharp but irregular border between the colors of the back and underside.

Adult Cascade Torrent Salamander

Adult Columbia Torrent Salamander

Adult Olympic Torrent Salamander

FULLY TERRESTRIAL

Ensatina

SALAMANDERS

CLOUDED SALAMANDER

Aneides ferreus COPE, 1869

Adults

STRONG, *fast and active, Clouded Salamanders are usually seen scrambling away or even jumping to a log or tree (other salamander species typically remain motionless when first discovered), so the one that got away from you may have been a Clouded Salamander. This species is quite hardy and versatile, and it appears to be able to withstand major disturbances to its forest habitat, particularly fires that remove entire stands of trees. Huge logs in forest openings are its typical habitat, but it has been found in urban settings where there is no residual forest nearby.* • Aneides *means 'shapeless,' probably in reference to the color pattern, which is random;* ferreus *means 'iron-colored,' and it refers to the rusty metallic color of the tail of many juveniles or to the pale metallic flecks on the adult.*

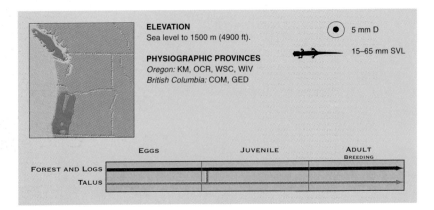

ELEVATION
Sea level to 1500 m (4900 ft).

PHYSIOGRAPHIC PROVINCES
Oregon: KM, OCR, WSC, WIV
British Columbia: COM, GED

● 5 mm D

15–65 mm SVL

	EGGS	JUVENILE	ADULT BREEDING
FOREST AND LOGS			
TALUS			

HABITATS: Clouded Salamanders live in forests, including burned, second-growth and rocky areas, primarily those with large-diameter logs (especially Douglas-fir). They lay their eggs in cavities in logs or in rock crevices. Juveniles and adults live in partially decayed logs, often those with wood that splits or breaks into layers and blocks. They are frequently found under the loose bark of logs or stumps, especially those with solid wood under the bark. During rainy periods they can be found under wood and bark debris on the forest floor, or occasionally in snags, and they also live in moist rock outcrops.

EGGS are rarely found. The female lays several of the white eggs together, and she stays with them until they hatch. The eggs are laid in a bunch that is attached to the roof of a cavity, with each egg supported by a jelly strand. The strands often become twisted together.

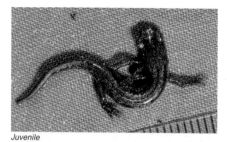

JUVENILES have a short body, and the toes overlap if the legs are adpressed. The large hind feet are as wide as they are long. The outer (fifth) toe on the hind foot is about as long as the fourth toe. Juveniles are dark gray, with a dense concentration of bright metallic flecks (usually bronze, brass or gold), which often form a solid color on the back. The color is especially bright on the tail, the tops of the leg bases and the snout (in front of the eyes).

Juvenile

Top of right hind foot

ADULTS have a body that is not very long, and the toes touch or may slightly overlap if the legs are adpressed. The toes are slightly wider near their tips than mid-length, and the tips are distinctively blunt or squared. The large hind feet are as wide as they are long. The outer (fifth) toe of the hind foot is about as long as the fourth toe. Adults are translucent gray or tan (occasionally almost black), with varying amounts of irregularly shaped metallic flecks. The flecks are pale gold, brass or bronze, and they are most concentrated on the tail, the tops of the leg bases and the snout (in front of the eyes). The underside is a paler gray than the back.

Camouflaged adult

BLACK SALAMANDER

Aneides flavipunctatus (STRAUCH, 1870)

Adult

THE *Black Salamander is mostly a Californian species, and its range barely enters southern Oregon. Not much is known about its habitat needs or where it lays its eggs in our region. The Black Salamander is the largest of our fully terrestrial salamanders. It is as strong and as active as the Clouded Salamander, but it is less likely to jump or run. Black Salamanders may be easier to catch, but they are more difficult to find, because one has to pick up lots of boulders and search underneath them.* • Aneides *means 'shapeless,' perhaps in reference to the random color pattern;* flavi·punctatus *means 'yellowish spotted,' and it refers to the yellowish- or greenish-white dots or flecks on the adults.*

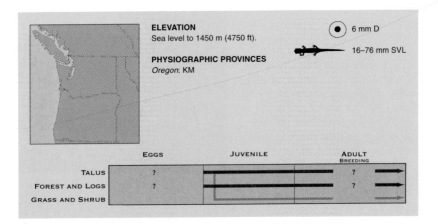

ELEVATION
Sea level to 1450 m (4750 ft).

PHYSIOGRAPHIC PROVINCES
Oregon: KM

6 mm D

16–76 mm SVL

	EGGS	JUVENILE	ADULT BREEDING
TALUS	?		?
FOREST AND LOGS	?		?
GRASS AND SHRUB			

HABITATS: Little is known of the Black Salamander's habitat requirements. It is found in forests, open woodlands, moist talus and streamside areas with logs and wood or rock debris. The one nest of eggs that has been documented was found underground in a cavity in the soil. Juveniles and adults live in logs and moist talus (more often under small boulders than under smaller rocks). They can also be found under surface debris during warm rainy periods.

EGGS have been documented from one nest. Several eggs were grouped together in a cavity in the soil. Each individual egg was attached by a strand of jelly to a central jelly base. A female was with the eggs.

Juvenile

JUVENILES have large, wide feet that are speckled pale pink, so it looks like the salamander has four pink hands. The toe tips are wide and rounded. The outer (fifth) toe is about three-quarters the length of the fourth toe. Juveniles are densely coated or speckled with tiny, bright metallic green or olive flecks on the head, back and tail, with a few larger white dots. The sides and belly have white or cream speckling and some larger white flecks.

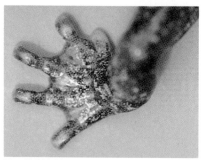
Top of left hind foot

ADULTS have a long body, and the toes do not touch if the legs are adpressed. The feet are large and almost as wide as they are long. They are speckled pink, so it looks like the salamander has four pink hands. The toes are slightly wider near their tips than at mid-length, and the tips are rounded. The outer (fifth) toe of the hind foot is about three-quarters the length of the fourth toe. Adults are black or very dark gray, and they are speckled (often sparsely) with tiny cream or greenish flecks on the head, back and tail. A few larger, whitish dots occur on the back, and there is a dense coating or speckling of whitish dots or flecks on the lower sides and on the belly, which is charcoal gray.

Adult

Adult

OREGON SLENDER SALAMANDER

Batrachoseps wrighti (BISHOP, 1937)

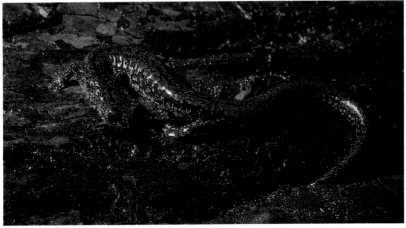

Adult

WHEN an Oregon Slender Salamander is uncovered, it usually coils up its body and tail (or just its body) and remains motionless, although it may flip about by suddenly uncoiling. There are many theories about the advantages of this coiling behavior, including that it may make the salamander look like a nasty-tasting millipede. Coiling may also help it spring or roll away—a fully coiled salamander can easily roll off a layer of wood you are investigating, and it may escape your notice. Once you have found ideal habitat conditions, you may find many individuals close together. This species is only common in stable, moist old-growth forests with many large, old logs. We have never found an Oregon Slender Salamander in an open clearcut, which makes us believe that we need to conserve natural forest conditions that include large logs, because this species does not appear to adapt quickly to severe change. • Batracho·seps *means 'frog-lizard,' perhaps in reference to its coiling and springing escape action;* wrighti *honors Dr. A.H. and Miss M.R. Wright, the herpetologist and his daughter who first discovered this species.*

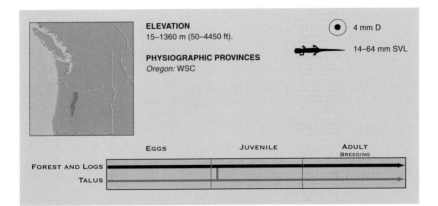

ELEVATION
15–1360 m (50–4450 ft).

PHYSIOGRAPHIC PROVINCES
Oregon: WSC

● 4 mm D

14–64 mm SVL

	EGGS	JUVENILE	ADULT BREEDING
FOREST AND LOGS			
TALUS			

HABITATS: Oregon Slender Salamanders live in forests, especially old-growth Douglas-fir and younger stands with large numbers of large logs. They lay their eggs under thick bark, inside a crevice in a log or in talus. Juveniles and adults live under thick bark, inside partially decayed logs or in debris piles around the bases of large snags. They are most often found in Douglas-fir wood that splits into clean, red layers or blocks (with little or no powdery, rotten wood). They also occur in moist talus that has abundant woody debris. During warm rainy periods, they can be found under woody or rock surface debris, usually on a clean slab of wood or leaves, rather than on soil or duff.

EGGS are not often found. The eggs are enclosed in a long, thin strand of jelly, which is wound around so that the eggs appear to be in a cluster. The female stays with the eggs.

Juvenile

JUVENILES have a long body and small legs and feet. The hind feet have only four toes. Juveniles are black or dark gray, with a mottled, dull reddish dorsal stripe, which becomes brighter with age.

ADULTS have a long, thin body, and the tail is as long as the head and body. The legs, feet and toes are tiny and delicate. The hind feet have only four short toes with slightly widened, round tips. Adults are black or dark gray, with a mottled dorsal stripe of rounded, gold-edged patches of brick or metallic red or orange. Irregular areas of dark body color show between patches of the stripe colors. Large white or silver flecks are scattered on the underside.

Hind feet

Coiled adult

Adult

CALIFORNIA SLENDER SALAMANDER

Batrachoseps attenuatus (ESCHSCHOLTZ, 1833)

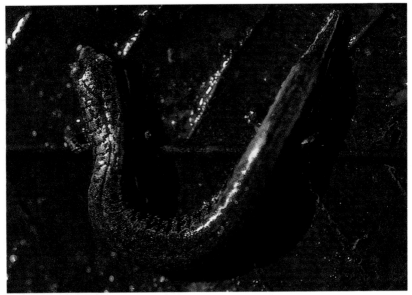

Adult with regenerating tail

THE *California Slender Salamander is only found in our region along the southern Oregon coast. Apparently, it can tolerate salt spray from the ocean. This species appears to be intolerant of open clearcuts, which remove large logs and allow moisture to evaporate from the site. Its primary habitat is in redwood forests, where its exquisitely delicate body and intricate color pattern contrast with the enormous logs in which it lives. Like the Oregon Slender Salamander, its colors provide perfect camouflage in the decaying wood it prefers, where its worm-like shape can most easily slide between layers of clean, red, rotting wood. When a California Slender Salamander is uncovered, it usually coils its body and tail (or just its body) and remains motionless. When fully coiled, it can easily roll off the layer of wood you are investigating, and it may escape your notice.* • Batracho·seps means 'frog-lizard,' perhaps in reference to its coiling and springing escape action; attenuatus means 'reduced' or 'weakened,' and it refers to this species' tiny legs.

ELEVATION
Sea level to 450 m (1500 ft).

PHYSIOGRAPHIC PROVINCES
Oregon: KM

4 mm D

12–47 mm SVL

	EGGS	JUVENILE	ADULT BREEDING
FOREST AND LOGS			

HABITATS: California Slender Salamanders live in humid coastal forests, especially among redwoods. They lay their eggs in logs, underground or under woody surface debris. Juveniles and adults live underground, in partially decayed logs or under bark or large logs. During rainy periods, they can be found on the forest floor under bark, other woody debris or rocks, usually on a clean slab of bark, wood or leaves, rather than on soil or duff.

Underside of adult

EGGS are not often found. The eggs are enclosed in a long, thin string of jelly, which is wound around so that the eggs appear to be in a cluster. The female apparently does not stay with the eggs.

JUVENILES have a long, thin body and tiny legs and feet. The hind feet have only four toes. Juveniles are brown or black, with a mottled dorsal stripe of pink, red and cream.

ADULTS have a body that is incredibly thin and fragile looking. The tail is longer than the head and body, but it is often broken off. The legs and feet are tiny; they look too small to be really useful. The hind feet have only four toes. The toes are short and round or oblong, so the feet look cat-like. Adults are brown or black, with a mottled dorsal stripe of small, round patches of pink, brick red and cream or tan. The stripe has a chevron design, with V's or zigzags of the dark body color in the gaps. Tiny white flecks are scattered on the underside.

Coiled adult

Adults

ENSATINA

Ensatina eschscholtzii GRAY, 1850

Adult

THE *Ensatina is a highly variable species that has many intergrading color forms (sub-species). Some geneticists believe this species is really a whole bunch of separate species. Ensatinas seems to be extremely adaptable to disturbances, and they can be found in every-thing from pristine old-growth forests to logging slash piles to urban wood scrap heaps. They spend considerable time underground, in rotting root tunnels and in the burrows of nightcrawlers and small rodents. • Ensat·ina means 'sword-like' or 'sword that is small,' and it refers to the way the tail is held straight and is reportedly brandished at predators; eschscholtzii honors J.F. Eschscholtz, a Russian naturalist who first described many species (though not the Ensatina) during an 1815 exploration of the northwest coast.*

ELEVATION
Sea level to 1400 m (4600 ft).

PHYSIOGRAPHIC PROVINCES
Oregon: ESC, KM, OCR,
    WSC, WIV
Washington: OP/SW, PT, WSC
British Columbia: COM, GED

4–7 mm D

17–65 mm SVL

	EGGS	JUVENILE	ADULT BREEDING
FOREST AND LOGS			
TALUS			

HABITATS: Ensatinas live in a variety of moist coniferous and deciduous forests and partially wooded habitats. They lay their eggs in a burrow, an underground cavity or in partially decayed logs. Juveniles and adults live in logs, debris piles at the bases of snags, moist talus that has abundant woody debris, and piles of firewood or old shingles or plywood. They are often found under surface debris during and after rainy periods. They live underground during dry or cold weather.

EGGS are rarely found. The eggs are whitish and they are laid in a cluster. The female stays with the eggs.

JUVENILES have a very short body and a large head. The constricted base of the tail may be hard to see. Juveniles are dark gray, brown or dull purple, with an almost luminescent yellow (or orange) on the upper surfaces of the legs, and they may have an orange tail. The head, back and sides are covered with angular, metallic-white or silver flecks, which may be arranged in large patches. There is a scattering of brassy flecks on the head and back.

Juvenile

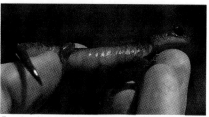

Female, with eggs visible inside

ADULTS have a very short body, and the toes overlap easily if the long legs are adpressed. The head and eyes are large. The tail is constricted at its base, where it may show several vague rings (best seen from the side), as if a rubber band had been left on for a while. Adults are variable in color. They are translucent red-brown, grading into pale tan or cream on the underside, with a pink tone overall. There is usually a bright yellow or orange patch on the upper surface of the base of the leg. In southern Oregon, adults often have a network of black markings over the translucent red-brown color. In southwest Curry County, Oregon, a small subspecies of Ensatina has black, orange and yellow spots and mottling that are brightest on the tail, and the underside is yellow or almost white.

Adult

Adult

DUNN'S SALAMANDER

Plethodon dunni BISHOP, 1934

Adult

THE largest Plethodon salamander in our region, Dunn's Salamander can be confused with the Long-toed Salamander, because their color patterns are somewhat similar, but their feet and habits are distinctly different. Salamanders of the genus Plethodon have small legs, and their feet have small toes. Although Dunn's Salamander lives next to streams and has slightly webbed feet, if it falls into the water it will struggle desperately to get out. When discovered, Dunn's Salamanders generally scramble down into the talus or duff.
• Pleth·odon means 'many teeth,' and it refers to the long row of teeth found in members of this genus; dunni honors E.R. Dunn, a herpetologist who studied salamanders in the 1920s and later.

ELEVATION
Sea level to 1000 m (3300 ft).

PHYSIOGRAPHIC PROVINCES
Oregon: ESC, KM,
    OCR, WSC, WIV
Washington: OP/SW

● 5 mm D

17–70 mm SVL

	EGGS	JUVENILE	ADULT BREEDING
MOIST TALUS			
EDGES OF MEDIUM STREAM	?		
EDGES OF SMALL STREAM			
EDGES OF LARGE STREAM			
STREAMSIDE FOREST AND LOGS			

HABITATS: Dunn's Salamanders live in the rocky edge of forested streams and permanently wet or moist talus. The one nest that has been documented was found in a rock crevice. Juveniles and adults live in gravel or under cobbles or small boulders at the edges of stream channels, on streambanks, in moist talus or in rock outcrops. They do not live in flowing water, but rather where the substrate remains moist. In rainy weather, they are occasionally found in or under logs near streams or under surface debris.

EGGS have been documented from only one nest. A cluster of eggs was attached by one jelly strand. The female was with the eggs.

JUVENILES have an outer (fifth) toe on the hind foot that is one-third to half the length of the fourth toe. Juveniles are black or dark brown, with a bright yellow (like hot dog mustard) or greenish-yellow dorsal stripe with a ragged edge. There are patches of the stripe color on the sides and on the upper surface of the entire length of the legs.

Juvenile

ADULTS have a long body, usually with 15 costal grooves, and the tail is about as long as the head and body (shorter in females). The legs are short, and if they are adpressed, the feet are separated by two to four intercostal folds. The hind foot appears slightly webbed when the toes are spread. The outer (fifth) toe on the hind foot is one-third to half the length of the fourth toe. Adults are dark brown-gray, usually with a wide dorsal stripe of dull mustard yellow (Dijon style), olive-tan or olive green. The edge of the stripe is distinct but ragged, and the stripe does not extend to the end of the tail (most noticeable on larger

Juvenile

Adult

individuals). Patches of the stripe color occur on the sides, as though pieces of stripe broke off and slid down onto the sides. The dorsal stripe color also usually occurs in small patches scattered on the upper surface of the entire length of the legs. The underside has whitish flecks.

LARCH MOUNTAIN SALAMANDER

Plethodon larselli BURNS, 1954

Adult

SCIENTIFIC knowledge of the Larch Mountain Salamander is just unfolding at this time. It was originally thought to be confined to the Columbia River Gorge (on both the Oregon and Washington sides), but its range has recently been extended far to the north. An old record from farther south in Oregon has never been confirmed. This species was also thought to be found only in talus slopes, but recent surveys near Mt. St. Helens, in Washington, have documented its occurrence in forest environments with no talus. Perhaps it is there that a nest and eggs of this species will first be found. When threatened, Larch Mountain Salamanders often coil up, and they may flip or spring away. • Pleth·odon means 'many teeth,' and it refers to the long row of teeth found in members of this genus; larselli honors Dr. O. Larsell, a former head of the Oregon Medical School.

ELEVATION
15–1050 m (50–3450 ft).

PHYSIOGRAPHIC PROVINCES
Oregon: ESC, WSC
Washington: ESC, WSC

● 4 mm D

17–53 mm SVL

	EGGS	JUVENILE	ADULT BREEDING
TALUS	?		?
FOREST AND LOGS	?		?

HABITATS: Larch Mountain Salamanders live in talus slopes in forested areas, usually away from streams, or on steep slopes in old-growth forests with abundant woody debris. No nests have ever been documented. Juveniles and adults can be found under surface rocks and woody debris during or after rainy periods, but they remain deep in the talus or woody debris during cold or dry weather. In Oregon, they are most often found in patches of gravel or small cobble-sized talus that is mixed with fine woody debris and little soil, and that has a thin or patchy layer of moss. In Washington, they are most often found in small cobble talus that has a heavy moss covering and is shaded by conifer trees, or they are found in old-growth forests, under surface debris or in debris piles around the bases of large snags.

EGGS have never been documented for this species.

JUVENILES have an outer (fifth) toe on the hind foot that is just a nubbin; it is one-quarter (or less) the length of the fourth toe. Juveniles are black or dark brown, with a golden-yellow to orange-red dorsal stripe. The stripe has ragged edges, and there are a few black marks in the center of the stripe, especially above the hind legs. A few flecks of the stripe color occur on the upper surface of the legs near the body. Along the sides there are tiny white flecks, which are largest on the sides of the belly. In the center of the belly there is a subtle yellow or salmon pink area that expands and brightens with age.

Juvenile

Small adult

ADULTS have a long body and short legs, and the tail is slightly shorter than the head and body. The hind foot appears slightly webbed when the toes are spread. The outer (fifth) toe on the hind foot is just a nubbin; it is less than one-quarter the length of the fourth toe. This salamander has a brick red, orange-brown or dull yellow dorsal stripe with ragged edges that extends to the end of the tail. The dark body color usually shows through the center of the stripe in irregular patches that form a thin black line. The sides are black, brown or gray, with zones of white flecks on the inter-costal folds. There may also be tiny flecks of the stripe color on the sides. A few flecks of the stripe color occur on the upper surface of the legs near the body. The underside is salmon pink (often quite bright), with white flecks on the sides of the belly.

Underside of adult

Adult

VAN DYKE'S SALAMANDER

Plethodon vandykei VAN DENBURGH, 1906

COEUR D'ALENE SALAMANDER

Plethodon idahoensis SLATER AND SLIP, 1940

Adult Coeur d'Alene Salamander

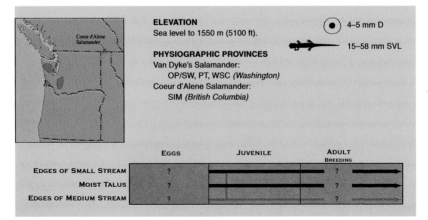

Adult Van Dyke's Salamander

MANY herpetologists believe that the Coeur d'Alene and Van Dyke's salamanders are actually one species that has been separated geographically into two populations that can no longer interbreed simply because of the distance between them. The known range of the Coeur d'Alene Salamander has recently been expanded into southeastern British Columbia, though it is still one of the less well known amphibians in our region. These species sometimes coil and occasionally flip away when discovered. • Pleth·odon *means 'many teeth,' and it refers to the long row of teeth found in members of this genus;* vandykei *honors E.C. Van Dyke, the first person to collect the species;* idaho·ensis *means 'belonging to Idaho.'*

ELEVATION
Sea level to 1550 m (5100 ft).

PHYSIOGRAPHIC PROVINCES
Van Dyke's Salamander:
OP/SW, PT, WSC *(Washington)*
Coeur d'Alene Salamander:
SIM *(British Columbia)*

4–5 mm D

15–58 mm SVL

	EGGS	JUVENILE	ADULT BREEDING	
EDGES OF SMALL STREAM	?		?	
MOIST TALUS	?		?	
EDGES OF MEDIUM STREAM	?		?	

HABITATS: Van Dyke's and Coeur d'Alene salamanders can live on streambanks, seeps or moist, north-facing, rocky habitats in forested areas. Two documented nests were found under a mossy stone and inside a log near a stream. Juveniles and adults live under cobble or woody debris in stream edges, in logs adjacent to streams and in seeps. They also occur on shaded or north-facing slopes in moist, mossy talus or fractured rock outcrops with slowly seeping water.

EGGS have been documented from only two nests. The cluster of eggs was attached by a single strand of jelly to a rock in one nest and to the inside of a log in the other. The female was with the eggs at one of the nests.

JUVENILES have a short body, and the toes touch if the legs are adpressed. Juveniles are black or dark gray, with a bright yellow or salmon pink-orange dorsal stripe. The throat has yellowish patches. Small patches of stripe color may occur on the upper surfaces of the legs near the body. There are white flecks on the sides.

ADULTS are shorter and stockier through the body and tail than other *Plethodon* salamanders. There are 14 costal grooves in total, and if the legs are adpressed, the feet are separated by one to three intercostal folds. The throat is yellow. The parotoid glands are visible at the back corners of the head. Each gland is a small, teardrop-shaped, rough or pitted lump, outlined by a slight groove. The hind feet have slight webbing, visible when the toes are spread. The outer (fifth) toe on the hind foot is about one-third the length of the fourth toe. • **Van Dyke's Salamander** has three color phases. The dark phase is blackish, with a yellow, reddish or greenish dorsal stripe and white flecks on the belly. The yellow phase is yellow-tan, with a slightly brighter stripe. The rose phase is dull salmon pink, with a slightly brighter stripe. • The **Coeur d'Alene Salamander** has only a dark color phase. The dorsal stripe is a similar color to that of Van Dyke's Salamander, but it has conspicuously ragged edges. It is narrow and breaks into patches on the head and the extreme tip of the tail. There may be no stripe color on the bases of the legs. The throat may have blotches of pale yellow.

Juvenile Coeur d'Alene Salamander

Juvenile Van Dyke's Salamander

Adult Van Dyke's Salamander, showing parotoid glands

Underside of adult Coeur d'Alene Salamander

WESTERN REDBACK SALAMANDER

Plethodon vehiculum (COOPER, 1860)

Adult

O NE of the smallest and most abundant of the woodland salamanders, the Western
Redback Salamander often belies its common name with a yellow or dull tan back; it
may even have a very dark back with no dorsal stripe. When its hiding place is discovered, it
is most likely to try to escape by wriggling down into the talus or rotten wood. • Pleth·odon
means 'many teeth,' and it refers to the long row of teeth found in members of this genus;
vehiculum seems like a very strange name for a salamander, because it means 'a vehicle' or
'something that carries something else.' It refers to the courtship behavior of this species (and
of most other Plethodon species), in which the male carries the female while she holds onto
his tail.

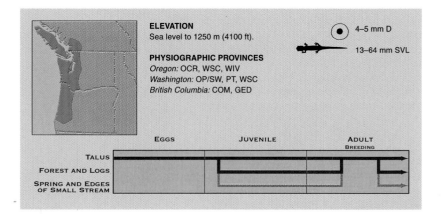

ELEVATION
Sea level to 1250 m (4100 ft).

PHYSIOGRAPHIC PROVINCES
Oregon: OCR, WSC, WIV
Washington: OP/SW, PT, WSC
British Columbia: COM, GED

4–5 mm D

13–64 mm SVL

	EGGS	JUVENILE	ADULT BREEDING
TALUS			
FOREST AND LOGS			
SPRING AND EDGES OF SMALL STREAM			

HABITATS: Western Redback Salamanders live in a variety of habitats in forested areas, often in talus. The eggs have been found hidden between rocks. Juveniles and adults live underground, or in logs that are in various stages of decay (but usually where the wood breaks into blocks or layers), or in shaded or slightly moist talus, and occasionally in springs and stream edges. During rainy periods, they are often found under surface rocks and woody debris. During cold or dry weather, they remain in talus, in large logs or underground.

Juvenile

EGGS are rarely found. They are laid in a cluster. The female stays with the eggs.

JUVENILES have an outer (fifth) toe on the hind foot that is about one-third the length of the fourth toe. Juveniles are variable in color, but they are usually black or dark gray, with white flecks on the sides and belly. The dorsal stripe (if present) is a broad, bright band of red or gold, and it may have a few irregular, dark patches. From above, you can see little or no dark color on the sides of the tail. The stripe color also occurs in a broad, sharp-edged band on the upper surface of the legs near the body. There may also be tiny flecks of stripe color on the sides.

Small adult

ADULTS have a long body with 16 costal grooves, and the tail is about as long as the head and body (shorter in females).

Hind feet

The legs are short, and if they are adpressed, the feet are separated by 2½ to 5½ intercostal folds. The outer (fifth) toe on the hind foot is about one-third the length of the fourth toe. Adults are variable in color. The background color is usually black or gray, but sometimes it is reddish or yellowish. There are irregular white flecks on the underside. The dorsal stripe (if present) is usually red or gold, but it is sometimes greenish or gray. The edges of the stripe are distinct and uniform (only slightly scalloped). The stripe is a broad, solid band from the hind legs to the tip of the tail; from above, you can see little or no dark color on the sides of the tail. From the hind legs forward, the stripe has irregularly scattered dark patches, and it fades out at the back of the head. The stripe color also occurs in a broad, sharp-edged band on the upper surfaces of the legs near the body. The sides may have tiny flecks of stripe color.

DEL NORTE SALAMANDER

Plethodon elongatus VAN DENBURGH, 1916

Adult

DEL NORTE is *Spanish for 'of the north,' but this reflects a Californian perspective, since the Del Norte Salamander only occurs in the extreme southwestern corner of our Pacific Northwest region. Although juveniles have a bright dorsal stripe, the dull, dark colors of the adults blend with the soil and duff. Del Norte Salamanders are often found under a rock or a piece of wood next to the entrance of a small burrow. They generally remain motionless when uncovered, but they will quickly slide down the burrow if they are touched.*
• Pleth·odon *means 'many teeth,' and it refers to the long row of teeth found in members of this genus;* elongatus *reflects that this species has the longest body proportions of any* Plethodon *salamanders in our region.*

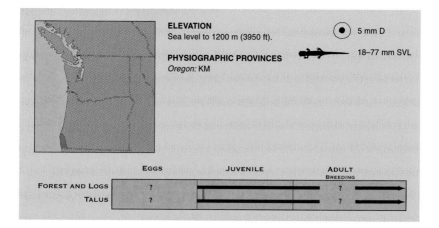

ELEVATION
Sea level to 1200 m (3950 ft).

PHYSIOGRAPHIC PROVINCES
Oregon: KM

5 mm D

18–77 mm SVL

	EGGS	JUVENILE	ADULT BREEDING
FOREST AND LOGS	?		?
TALUS	?		?

HABITATS: Del Norte Salamanders live in areas of coniferous or deciduous forest with rock or logs. The one nest that has been documented was found in a cavity under a post. Juveniles and adults live in moist talus or fractured rock outcrops, or in partially decayed logs (most commonly shaded by old-growth conifers or on north-facing slopes) that are usually away from streams. During rainy periods they can be found under surface rocks and wood debris. They remain underground or deep in talus during dry or cold weather.

EGGS have been documented from only one nest. A cluster of eggs was attached to the roof of the cavity by one jelly strand. The female was with the eggs.

JUVENILES are usually black, with a narrow, sharp-edged, bright red or orange dorsal stripe. The stripe breaks into patches on the outer quarter of the tail, and on the back it has a central zone or double line of tiny, dark speckling that increases with age. The underside may have a few light flecks. The body and legs appear black, and they contrast sharply with the dorsal stripe, but there may be tiny flecks of stripe color on the sides and on the upper surface of the legs near the body.

Juvenile

ADULTS have a very long and quite slender body with 17 to 20 costal grooves, and the tail is about as long as the head and body (shorter in females). The legs are short, and the feet are separated by $6^{1}/_2$ to $7^{1}/_2$ intercostal folds if the legs are adpressed. The outer (fifth) toe on the hind foot is about one-third the length of the fourth toe. Adults are medium to dark brown or gray. The dorsal stripe is vague or dull, and it is composed of round, brown or orange dots on a dark background. The throat may have a few light patches, and there may be tiny flecks of stripe color on the sides and the upper surface of the legs near the body.

Juvenile

Adult

Underside of adult

SISKIYOU MOUNTAINS SALAMANDER

Plethodon stormi HIGHTON AND BRAME, 1965

Adult

THIS species is quite similar to the Del Norte Salamander, and both occur only in the southwestern part of our region. The Siskiyou Mountains Salamander has so far only been found in or very close to talus slopes and rock outcrops. Like the Larch Mountain Salamander, its eggs have never been found, and they are most likely hidden deep in the rocks.
• Pleth·odon means 'many teeth,' and it refers to the long row of teeth found in members of this genus; stormi honors R.M. Storm, emeritus professor from Oregon State University, whose enthusiastic support has welcomed many newcomers to the field of herpetology. He is among the best salamander-chasing rock-rollers in the region.

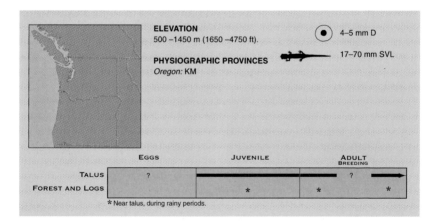

ELEVATION
500–1450 m (1650–4750 ft).

PHYSIOGRAPHIC PROVINCES
Oregon: KM

4–5 mm D

17–70 mm SVL

	EGGS	JUVENILE	ADULT BREEDING	
TALUS	?		?	
FOREST AND LOGS		*	*	*

* Near talus, during rainy periods.

HABITATS: Siskiyou Mountains Salamanders live in deep talus areas and fractured rock outcrops, especially on forested, north-facing slopes. No nests have been documented for this species. Juveniles and adults live in talus and broken rocks. During rainy periods, they may also be found under woody debris near talus slopes.

Juvenile

EGGS: No nests have been documented for this species.

JUVENILES are black, with a sparse to heavy sprinkling of white flecks, especially on the head and sides. The dorsal stripe is a zone of dull gold flecks. Its edges are straight, but they are not sharply defined, and the stripe fades out on the head and tail.

Juvenile

ADULTS have a very long body with 17 costal grooves, and the tail is about as long as the head and body (shorter in females). The legs are short, and the feet are separated by 4 to 5$\frac{1}{2}$ intercostal folds if the legs are adpressed. The outer (fifth) toe on the hind foot is about one-third the length of the fourth toe. Adults are pink-tan or pink-gray to light brown. There is a sparse to generous sprinkling of white flecks or small patches, especially on the head and sides. The dorsal stripe (if present) is vague, and it is composed of pinkish or golden-tan dots.

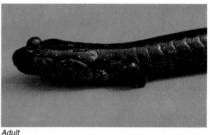
Adult

Female with eggs visible inside

FROGS & TOADS

Cascades Frog

TAILED FROG

Ascaphus truei STEJNEGER, 1899

Adult

WHILE *most tadpoles and frogs seek warm water or bask in the sun, the Tailed Frog is strongly adapted to cold conditions in our region. It occurs in very cold, fast-flowing streams that are often darkly shaded. The Tailed Frog is the only amphibian species in the Pacific Northwest to have internal fertilization of the eggs, which prevents the sperm from being washed away. Hatchlings are striking because they have no pigment and are almost transparent. Tailed Frogs develop very slowly in the cold water, and tadpoles are two to five years old before they metamorphose. Juveniles take another few years to reach sexual maturity. Adults occur in a variety of colors, and they are often well camouflaged in the gravel and moss. • A·scaphus means 'no shovel' or 'no digging,' reflecting this species' non-digging habits and its smooth feet, which do not have bumps or blades for digging; truei honors F.W. True, who was head curator of the biology department of the U.S. National Museum.*

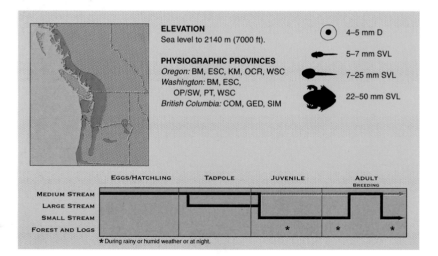

ELEVATION
Sea level to 2140 m (7000 ft).

PHYSIOGRAPHIC PROVINCES
Oregon: BM, ESC, KM, OCR, WSC
Washington: BM, ESC,
OP/SW, PT, WSC
British Columbia: COM, GED, SIM

4–5 mm D

5–7 mm SVL

7–25 mm SVL

22–50 mm SVL

	EGGS/HATCHLING	TADPOLE	JUVENILE	ADULT BREEDING
MEDIUM STREAM				
LARGE STREAM				
SMALL STREAM				
FOREST AND LOGS			*	* *

★ During rainy or humid weather or at night.

HABITATS: Tailed Frogs live in fast, small, permanent forest streams with clear, cold water, cobble or boulder substrates, and little silt. They lay their eggs in streams in summer, attaching them under cobbles or boulder-sized rocks. Tadpoles cling to the undersides (or less commonly the tops) of moss-free small boulders or large cobbles. They are more likely to be found lower in a stream than froglets and adults, which occur in gravel or under large cobbles, often in very shallow water. Adults often occur on streambanks at night and during wet weather. They may be found away from streams during winter rains and occasionally even on warm, humid, cloudy days.

EGGS are rarely found. They are laid in strings. Individual eggs are large and have no color.

HATCHLINGS have a conspicuous, huge, yellow gut full of yolk that is easily visible through the clear skin, which gradually develops gray speckling. The mouth is large, and it is surrounded by a cup-shaped band used for suction to rocks in fast-flowing streams. There are no external gills.

Hatchling

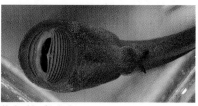

TADPOLES have a wide (top view) and flattened body. The underside looks very flat (side view). The huge mouth opens down, and it is surrounded by a wide, flat band used for suction to rocks in fast-flowing streams. It has two to three tooth rows on top and eight to 13 on the bottom. The dorsal fin is low, its top edge is tapered or almost straight, and it is opaque and as dark as the body. Tadpoles are usually black or reddish brown, often with a white spot at the tip of the tail.

Underside of tadpole

JUVENILES AND ADULTS have a small body, long legs and a large head. The eyes have vertical pupils. The outer toes on the hind feet are flat and wide, especially the outermost (fifth) toe. Males have a short, wide 'tail' that they use to internally fertilize the females. Frogs are usually tan or brown (sometimes green or red), often with indistinct dark blotches, and usually with a light bar or triangle between the eyes and snout. The skin may be grainy. They have no voice.

Juvenile

Underside of left hind foot

GREAT BASIN SPADEFOOT

Spea intermontana (COPE, 1883)

ALSO CALLED *SCAPHIOPUS INTERMONTANUS*

Adult

THE *Great Basin Spadefoot is a champion digger—it can disappear from sight in a few minutes. It uses alternating circular motions of its hind feet to push sandy dirt out from under itself and eventually up over its sides and back. Great Basin Spadefoots normally come out at night to catch insects, particularly after rains. Spadefoots may remain inactive underground until conditions are favorable. There are almost mythological anecdotes about spadefoots not being seen during a long and severe drought, until in the seventh year it finally rains and they come back up out of the ground and breed. Spadefoot tadpoles eat almost anything, including each other on occasion. • Spea means 'cave' and Scaphiopus means 'digger' or 'shovel,' and both refer to this species' habit of digging into the soil; inter·montana means 'between the mountains,' in reference to the dry lands between the Rocky Mountains and the Cascades or Coast Mountains.*

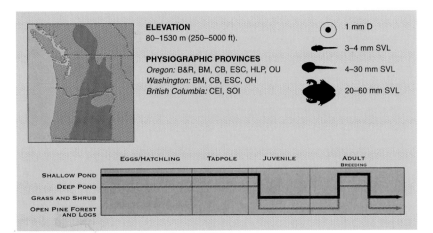

ELEVATION
80–1530 m (250–5000 ft).

PHYSIOGRAPHIC PROVINCES
Oregon: B&R, BM, CB, ESC, HLP, OU
Washington: BM, CB, ESC, OH
British Columbia: CEI, SOI

1 mm D

3–4 mm SVL

4–30 mm SVL

20–60 mm SVL

	EGGS/HATCHLING	TADPOLE	JUVENILE	ADULT BREEDING
SHALLOW POND				
DEEP POND				
GRASS AND SHRUB				
OPEN PINE FOREST AND LOGS				

HABITATS: Great Basin Spadefoots live in semi-arid areas of sagebrush, grassland or open forest with sandy soil. They breed in pond edges, stock tanks or rain-filled depressions. Spadefoots can take advantage of a variety of ponds, including cow wallows, that dry up in a few weeks. After rains in mid to late spring, they lay their eggs in water less than 0.5 m deep. They loosely attach the clusters to submerged vegetation or lay them on the bottom. Tadpoles can live in water that is as warm as 90° F, including very dirty water. Juveniles and adults dig down into sandy soils, and they can be found on the surface at night during humid or rainy weather.

EGGS are in soft, randomly shaped, grape- to plum-sized clusters. Individual eggs are tan-gray above and cream below, and you can easily detach one from the group. In warm water, they hatch in two or three days.

HATCHLINGS have a broadly triangular head (top view) that appears to be distinctly separated from the belly. There is a prominent, narrow ridge connecting the head to the tail. Hatchlings are light tan or gray. They develop very quickly, and the tiny gills are often covered within a few hours of hatching.

Egg clusters

TADPOLES have a triangular head that is wider than the belly and appears to be separated from the body (top view) when they are small (approximately 4–20 mm SVL). Larger tadpoles have a somewhat flattened back. The spiracle is lower on the side than in other species and touches the belly outline (side view). The eyes are close together, and they are oriented somewhat upward. The nostrils are prominent and close to the eyes. The mouth is wide, and it has two to five tooth rows on top and three to four on the bottom. Tadpoles are dark gray-brown, with gold or brassy flecks or patches scattered all over and forming a conspicuous 'Y' across the body.

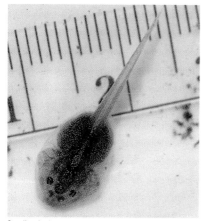

Hatchling

Small tadpole

Large tadpole

JUVENILES AND ADULTS have a wide, flat body and short legs. The eyes, which are separated by a hard lump, are large and have vertical pupils. The nostrils are raised, which gives this toad a pug-nosed profile. The hind feet have a single, black, hard, spade-shaped knob on the heel (for digging). Spadefoots are gray or tan, often with irregular, vague stripes and blotches. The skin has scattered, small, round bumps that are often red. The call is a loud, long, low-pitched, nasal quacking, like a slowed-down recording of ducks.

Juvenile

Underside of right hind foot

Half-buried adult

WESTERN TOAD

Bufo boreas BAIRD AND GIRARD, 1852

Adult

THE *Western Toad's camouflage coloring and skin texture, and its ability to bury itself in the dirt, help it avoid predators. If a Western Toad is caught, it may twitter (a vibrating sound also used to say 'hands off' to male toads during breeding), grumble, puff up to look bigger, urinate profusely or secrete a bitter poison. Western Toad tadpoles often move slowly through the shallows in huge, sinuous swarms that may stretch for 100 m. In late summer, uncountable hordes of tiny toadlets leave the water. If the weather turns cold, the toadlets cluster in deep piles. Increased exposure to ultraviolet radiation and the spread of an egg fungus are among the many theories to explain the observed reductions in Western Toad populations in many areas.* • Bufo *is Latin for 'toad';* boreas *is a Greek name for the north wind.*

ELEVATION
Sea level to 2250 m (7400 ft).

PHYSIOGRAPHIC PROVINCES
Oregon: All provinces
Washington: All provinces
British Columbia: BOP, CEI, COM, GED, NBM, SOI, SIM, SBI

1.5 mm D

3.5–5 mm SVL

5–18 mm SVL

11–145 mm SVL

	EGGS/HATCHLING	TADPOLE	JUVENILE	ADULT BREEDING
SHALLOW POND				
DEEP POND				
MARSH				
FOREST AND LOGS				
GRASS AND SHRUB				
EDGES OF SMALL STREAM				
EDGES OF MEDIUM STREAM				
EGDES OF LARGE STREAM				

HABITATS: Western Toads live in a variety of forested, brush and mountain meadow areas. They breed in ponds or shallow lake edges in mid spring, and they lay their eggs on the bottom in water less than 0.5 m deep. Hatchlings and tadpoles live in the warmest, shallowest water available. Toadlets often live under rocks near ponds or in brush. Adults live underground, under large debris and in grass and brush. They may occur in streams or springs during dry periods (especially east of the Cascade Range).

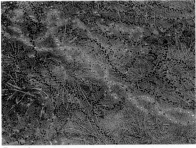

Egg strings

EGGS are evenly spaced in single file in very long, thin strings of jelly. Many females commonly lay their eggs loosely intertwined, creating a mat that covers an extensive area. Each string has two jelly layers, which may be visible with backlighting. Individual eggs are black above and white below.

HATCHLINGS have a vent that opens straight back, so the body is symmetrical. The tail is very short, with a low, short dorsal fin that is dark gray and translucent. Hatchlings are black or very dark maroon, and they may appear grainy.

TADPOLES have a square snout that juts forward from the round body outline. The mouth has two tooth rows on top and three on the bottom. The vent opens straight back. The tail is not much longer than the body. The dorsal fin is low, and it starts at the base of the tail trunk. Tadpoles are black or charcoal. The underside may be slightly paler, but the tail trunk is uniformly dark. The dorsal fin is dark, translucent and densely speckled gray or black.

Hatchlings

Tadpole

JUVENILES have a wide body and very short legs. They have dry, bumpy skin, and the belly is dark gray, with light flecks. The undersides of the feet have scattered, orange bumps.

ADULTS have a wide body and short legs. The parotoid glands are oval, and they are prominent. The hind feet have two large, round, pale, rubbery knobs on the heel for digging. Adults are a variety of colors, with a cream background covered with blotches of brown, gray, green or red. There is usually a thin, pale green or cream stripe down the back. The skin is dry and bumpy. The call is a quiet, bird-like twittering.

Juvenile

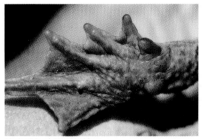

Underside of right hind foot

Adult

Clustered toadlets

87

WOODHOUSE'S TOAD

Bufo woodhousii GIRARD, 1854

Adult

WOODHOUSE'S Toad *is quite similar to the Western Toad. It occurs only along a few river valleys here at the western edge of its range. It is no longer found at some of its formerly recorded sites. Although the large-scale irrigation of croplands can create suitable wetland habitat, it can also reduce water quality where there is a high evaporation rate that concentrates agricultural chemicals. This species survives the very hot summers by burying itself. It uses its powerful hind feet, which are equipped with two small, hard knobs that loosen the dirt and may cut rootlets. Could its digging aerate the soil and improve plant growth? Woodhouse's Toads come out at night to catch insects, especially after rain or if lawn sprinklers are turned on. • Bufo is Latin for 'toad'; woodhousii honors S.W. Woodhouse, an army surgeon and naturalist in the Southwest in the mid 1800s.*

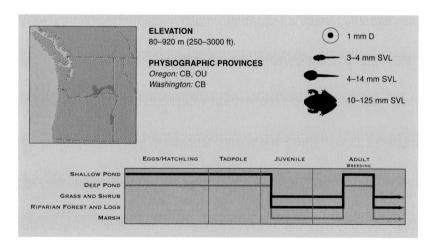

ELEVATION
80–920 m (250–3000 ft).

PHYSIOGRAPHIC PROVINCES
Oregon: CB, OU
Washington: CB

● 1 mm D

3–4 mm SVL

4–14 mm SVL

10–125 mm SVL

	EGGS/HATCHLING	TADPOLE	JUVENILE	ADULT BREEDING
SHALLOW POND				
DEEP POND				
GRASS AND SHRUB				
RIPARIAN FOREST AND LOGS				
MARSH				

HABITATS: Woodhouse's Toads live in river valleys in sagebrush or grassland areas. They breed in ponds, lakes, slow streams or irrigation ditches in late spring, and they lay their eggs in warm, shallow water. Hatchlings and tadpoles live in the warmest water available. Juveniles and adults usually live under debris or in dense brush. They stay underground in dry weather, and they can be found at the surface at night during or after rain.

EGGS are evenly spaced in single file in very long, thin strings of jelly. Many females commonly lay their eggs loosely intertwined, creating a mat that covers an extensive area. Each string has a single jelly layer, which may be visible with backlighting. Individual eggs are black above and white below. Woodhouse's Toad eggs are difficult to distinguish from Western Toad eggs.

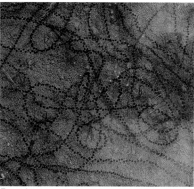

Egg strings

HATCHLINGS have a wide head. The vent opens straight back, so the body is symmetrical. Hatchlings are generally black or charcoal, but they are lighter on the belly and the underside of the tail trunk. The dorsal fin is colorless or pale gray.

Hatchling

TADPOLES have a round body with a jutting snout. The mouth has two tooth rows on top and three on the bottom. The vent opens straight back. The low, colorless dorsal fin starts at the base of the tail. Tadpoles are dark gray or brown, with pale gold and cream markings. Small tadpoles have small gold flecks on the underside and near the base of the tail (which gives them 'gold butts'). The number and size of the flecks increase with age, and large tadpoles have large, pale gold and cream blotches over much of the dark underside and sides. The underside of the tail trunk is white or cream.

Large tadpole

Large tadpole

JUVENILES have a wide body and very short legs. They have dry, bumpy skin and a whitish belly. The underside of the feet have scattered whitish bumps.

ADULTS have a very wide body, short legs and a very blunt snout. There are ridges between and behind the eyes; they are shaped like back-to-back L's. The parotoid glands are long ovals. The hind feet have two pale, hard, rubbery knobs on the heel, one spade-shaped and the other round. Adults have a cream background color covered with irregular grayish or brownish markings, and often with a thin cream line down the middle of the back. The skin is dry and bumpy. The call is a loud, long, shrill whistle that stays on one pitch.

Juvenile with tail remnant

Adult, showing ridges

Underside of left hind foot

PACIFIC TREEFROG

Hyla regilla BAIRD AND GIRARD, 1852

ALSO CALLED THE PACIFIC CHORUS FROG (*PSEUDACRIS REGILLA*)

Adult

THE Pacific Treefrog is the smallest, most common and most varied in color of the frogs of Oregon and Washington. It is often found far from water, and it can stay moist because its skin has glands that produce a waxy coating, which the frogs periodically eat. If you hear frogs in the Pacific Northwest, chances are good that you are hearing male Pacific Treefrogs, because most other species have much quieter calls and do not sing over such a long period of the spring. Studies have shown that one male acts as chorus master, leading the others to begin calling. Recordings of Pacific Treefrogs are used in jungle scenes in scores of Hollywood movies. It has been introduced to the Queen Charlotte Islands, B.C. • Hyla means 'woods' or 'bushes,' which reflects its typical habitat; regilla means 'regal' or 'splendid.'

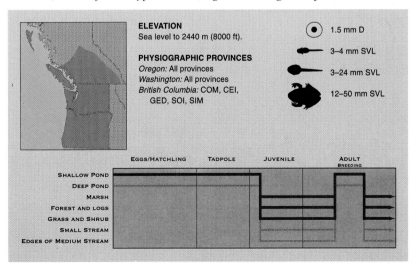

ELEVATION
Sea level to 2440 m (8000 ft).

PHYSIOGRAPHIC PROVINCES
Oregon: All provinces
Washington: All provinces
British Columbia: COM, CEI,
     GED, SOI, SIM

1.5 mm D

3–4 mm SVL

3–24 mm SVL

12–50 mm SVL

	EGGS/HATCHLING	TADPOLE	JUVENILE	ADULT BREEDING
SHALLOW POND				
DEEP POND				
MARSH				
FOREST AND LOGS				
GRASS AND SHRUB				
SMALL STREAM				
EDGES OF MEDIUM STREAM				

HABITATS: Pacific Treefrogs live in marshes, mountain meadows, woodlands, brush and disturbed areas. They breed in shallow ponds, seasonal pools, stock tanks or slow streams in early spring to early summer, and they readily move into new pools that last a short time. They attach their eggs to submerged grasses or twigs or lay them on the bottom, in water 0.5 m deep or less. Tadpoles live in the shallowest, warmest water available. Froglets live in grass or brush or under debris around ponds. Adults live in wet meadows or riparian areas, or well away from water in brush or woods. During dry weather east of the Cascade Range, they may remain underground (in burrows made by other animals) or in streams or springs.

EGGS are in a soft, tight, rounded mass (up to 4 cm in diameter). Each small egg has only a thin layer of jelly, so the eggs are packed closely together within the cluster. Individual eggs are tan to gray-brown above and yellow-gold or cream below.

Egg mass

HATCHLINGS are golden tan and tiny. The yolk is easily visible in the belly, which is at least as long and wide as the head. The eyes, which develop early, are conspicuously dark, and they modify the rounded body outline (top view). The top edge of the dorsal fin angles up from the middle of the body (side view). The gills are often no longer visible when hatchlings emerge from their eggs.

Hatchling

TADPOLES have a very short, round body. The belly is longer and wider than the head. The eyes poke out at the edge of the head (top view), modifying the rounded body outline. The snout is small and square, and the mouth has two tooth rows on top and three on the bottom. The tail is short. The top edge of the dorsal fin angles up from the middle of the back, but it levels out quickly, so the dorsal fin is no taller than the thickness of the tail trunk at its base (side view). The ventral fin has a straight lower edge. The tadpole may be 10 mm SVL when the hind leg buds appear, and it may metamorphose at 20 mm SVL. Small tadpoles (approximately 3–10 mm SVL) have dark, ragged stripes along their golden-tan bodies. Larger tadpoles are various colors; they are often mottled greenish or tan, with metallic patches and a blue sheen on the sides. The belly is usually white or silver.

Tadpole

Large tadpole

JUVENILES AND ADULTS are small. The toes are long and straight, and they do not taper to points. Their tips have round toe pads (wider than the toes, but hard to see on juveniles), that are used for clinging to smooth surfaces. Frogs are extremely varied in color. They are frequently bright green or pale tan, often with long, dark blotches. A sharply defined black mask extends in a line from the tip of the snout to the shoulder. Males have a dark gray, inflatable throat pouch during the breeding season, and they give loud 'ribbet' calls and choruses of loud, short, high-pitched trills. They also give a low 'krrreck' call during humid or rainy periods at any time of year.

Adult male calling

Underside of left hind foot

Juvenile

Adult

STRIPED CHORUS FROG

Pseudacris triseriata (WIED-NEUWIED, 1838)

ALSO CALLED THE BOREAL CHORUS FROG (*PSEUDACRIS MACULATA*)

Adult

ALTHOUGH *the Striped Chorus Frog occurs over a broad range elsewhere, in this region it has only been found in the northeastern corner of British Columbia. It is our smallest frog. Striped Chorus Frogs are often confused with Pacific Treefrogs. The males of both species have a balloon-like throat pouch that makes their calls louder. It is difficult enough to find a tiny frog hiding in a marsh, but when the male Chorus Frog calls, its voice sounds like it is coming from someplace else. You may hear one, but you are a good frogger if you can see one.* • Pseud·acris *means 'false locust,' probably in reference to its call, which can sound like a cricket or locust;* tri·seriata *means 'three striped'—the color markings are often arranged in three lines along the back.*

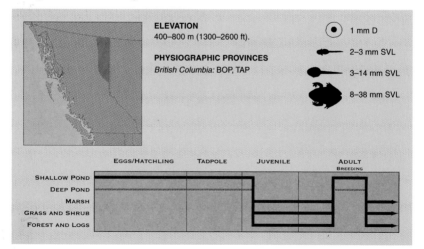

ELEVATION
400–800 m (1300–2600 ft).

PHYSIOGRAPHIC PROVINCES
British Columbia: BOP, TAP

◉ 1 mm D

2–3 mm SVL

3–14 mm SVL

8–38 mm SVL

	EGGS/HATCHLING	TADPOLE	JUVENILE	ADULT BREEDING
SHALLOW POND				
DEEP POND				
MARSH				
GRASS AND SHRUB				
FOREST AND LOGS				

HABITATS: Striped Chorus Frogs live in marshes, wet meadows, moist woodlands or brush. They breed in seasonal pools, flooded meadows, shallow pond or lake edges or the slow parts of streams, in early spring to early summer. They attach their eggs to emergent vegetation in water 0.5 m deep or less. Tadpoles live in the shallowest, warmest water available. Juveniles and adults live in wet meadows, riparian areas or moist brush or woods. Adults may spend dry periods under leaves at the base of willows or brush, or often in a burrow underground.

Egg mass

EGGS are in a soft, long cluster (usually less than 2.5 cm in diameter) that encloses a submerged grass blade or soft stem. They are similar to Pacific Treefrog eggs, but individual Striped Chorus Frog eggs are smaller, and the egg cluster is longer. The eggs are densely packed within the cluster.

Eggs about to hatch

HATCHLINGS are very tiny and gray in color. They have dark, prominent adhesive glands and eyes. The yolk is easily visible in the round belly, which is shorter and narrower than the head. The tail fins are low and straight.

Hatchling

TADPOLES are very similar to Pacific Treefrog tadpoles, but smaller at every stage. Striped Chorus Frog tadpoles may be only 5 mm SVL when the hind leg buds appear, and they may metamorphose at only 12 mm SVL. The belly is shorter and narrower than the head (top view). The nostrils are large and prominent. The eyes protrude slightly from the edge of the body (top view). The mouth has two tooth rows on top and three on the bottom. They are gray or brown, speckled with gold flecks.

Tadpole

Tadpole

JUVENILES AND ADULTS are tiny. The hind legs are short (the lower part of the leg is about one-third of the snout-to-vent length), and the thighs are noticeably shorter than the lower legs. The webbing on the hind feet is merely a narrow strip along the bases of the toes. The toes are long and straight, not tapered, and the toe pads appear separate and are the same width as the toes

Juvenile

or narrower. Frogs are tan, gray or green, with olive, dark green or brown blotches or irregular stripes along the back. They have a sharply defined black mask that extends in a stripe from the tip of the snout across the shoulder to the side or to the groin. Their skin is grainy, especially on the underside. Males have a greenish-yellow or gray-green, inflatable throat during the breeding season. They give high-pitched, rising trills that are similar to the sound of a finger running over the teeth of a comb.

Underside of adult

Adult male calling

Chorus of males

RED-LEGGED FROG

Rana aurora BAIRD AND GIRARD, 1852

Adult

THE *Red-legged Frog is the big frog of Pacific Northwest forests in the lowlands west of the Cascades. It prefers cool conditions, particularly for breeding. Leaving shade trees around ponds and slow streams, and managing runoff from roads, are at least a beginning towards protecting the health of this species. To catch a Red-legged Frog (and most other frogs) requires a slow, sneaking approach, because you cannot run through the brush as fast as frogs can leap underneath it.* • *Rana is Latin for 'frog'; aurora means 'dawn,' and it probably refers to the pink and red colors of the legs, which are like the pink sky at dawn.*

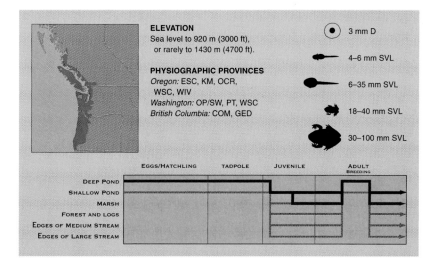

ELEVATION
Sea level to 920 m (3000 ft),
or rarely to 1430 m (4700 ft).

PHYSIOGRAPHIC PROVINCES
Oregon: ESC, KM, OCR,
WSC, WIV
Washington: OP/SW, PT, WSC
British Columbia: COM, GED

3 mm D
4–6 mm SVL
6–35 mm SVL
18–40 mm SVL
30–100 mm SVL

EGGS/HATCHLING · TADPOLE · JUVENILE · ADULT BREEDING

DEEP POND
SHALLOW POND
MARSH
FOREST AND LOGS
EDGES OF MEDIUM STREAM
EDGES OF LARGE STREAM

Egg masses

Hatchling

Tadpole

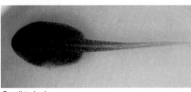
Small tadpole

HABITATS: Red-legged Frogs live in moist coniferous or deciduous forests and forested wetlands. They breed in water 0.5–2 m deep (occasionally deeper), in cool, usually well-shaded ponds or lake edges, beaver ponds or slow streams, in winter to early spring. They usually attach their eggs to a submerged branch or stem, or lay them on bottom vegetation. Hatchlings cling to the egg mass or to nearby vegetation. Tadpoles live in the warmer parts of a pond. During summer, froglets and adults live along streams, in moist sedge or brush, on shaded pond edges or under logs or debris. During damp conditions, they may occur in forests far from water.

EGGS are in a soft, grapefruit- to cantaloupe-sized mass. Egg masses may be close together, but they are not laid on top of each other. Individual eggs are black above and white below. They are very large, and each has a wide layer of jelly. Before hatching, the egg mass often floats to the surface, where it spreads out and may look frothy.

HATCHLINGS are stubby looking. The gills are very short nubbins. The tail angles up slightly from the body, and it is about one to $1\frac{1}{2}$ times the body length. The top edge of the dorsal fin arches up from the middle of the back, and the fin is tall and translucent light gray. (Also see Confusing Species Comparisons, p. 149.)

TADPOLES look stubby, because the tail is $1\frac{1}{2}$ times the body length or less, and because the dorsal fin is taller than the thickness of the tail trunk near its base (side view). The top edge of the dorsal fin arches steeply up from the middle of the back (above or in front of the spiracle). Just before metamorphosis, tadpoles may be quite large, with the tail more than $1\frac{1}{2}$ times the body length. The mouth has three tooth rows on top and four on the bottom. Tadpoles have a vague gold line along each side of the back when they are small (approximately 8–15 mm SVL). Larger tadpoles are tan, with bright gold or brassy blotches, especially on the underside. These blotches are smaller on the back. The dorsal fin has gold and light dots, and it often has a golden tone. (Also see Confusing Species Comparisons, p. 150.)

Hind feet

JUVENILES have gold eyes that look out to the sides. The dorsolateral folds are distinct (they may be broken or less distinct on the lower back), and they are well separated, so the lower back appears wide and rounded between them. The underside of the thigh is pale pink. The lip line is short, absent or blurred on the snout. (Also see Confusing Species Comparisons, p. 151.)

ADULTS have gold eyes that look to the sides. The hind legs are long, and the webbing is stepped down along the inner edges of the toes. There are bold cream to yellow and black (or red) patches in the groin that are larger than the color patches further forward on the side. The underside of the hind leg is translucent red; it looks like you can see red muscle under the skin. The mask and lip line are usually vague. The back may be unmarked, or it may have black speckling or very irregular black marks. The call is a quiet, low-pitched, throaty or muffled stuttering. (Also see Confusing Species Comparisons, p. 152.)

Juvenile

Underside of adult

Leaping juvenile

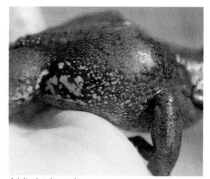
Adult, showing groin

CASCADES FROG

Rana cascadae SLATER, 1939

Adult

CASCADES Frogs lay their eggs in sunny ponds at high elevations. Recent studies indicate that increased levels of ultraviolet light may be responsible for a decrease in the number of eggs that hatch. Studies are continuing to examine the relationship between the depletion of the ozone layer by pollution, increasing ultraviolet light levels, and hatching success. Although declines in some local populations have been noted, the Cascades Frog is still a common species within its range. Studies have shown that tadpoles may recognize and stay with the other tadpoles from their egg mass. • Rana is Latin for 'frog'; cascadae refers to the Cascade Mountains, where this species lives, as well as to the cascading streams for which the mountains were named.

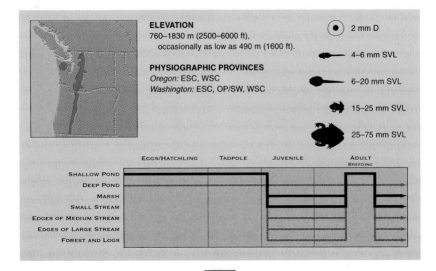

ELEVATION
760–1830 m (2500–6000 ft),
    occasionally as low as 490 m (1600 ft).

PHYSIOGRAPHIC PROVINCES
Oregon: ESC, WSC
Washington: ESC, OP/SW, WSC

2 mm D

4–6 mm SVL

6–20 mm SVL

15–25 mm SVL

25–75 mm SVL

	EGGS/HATCHLING	TADPOLE	JUVENILE	ADULT BREEDING
SHALLOW POND				
DEEP POND				
MARSH				
SMALL STREAM				
EDGES OF MEDIUM STREAM				
EDGES OF LARGE STREAM				
FOREST AND LOGS				

HABITATS: Cascades Frogs live in mountain meadows and moist forests. They breed in bogs or ponds with cold springs or snowmelt in early to mid spring, and they pile their eggs on barely submerged mosses or other short vegetation, in water less than 20 cm deep. Hatchlings may be stranded in the moisture on top of the egg mass at first; then they swim in rain or snowmelt to deeper water. Tadpoles live in the warmest parts of a pond. Froglets and adults live in wet meadows, bogs or brush, or along forested stream or pond edges in summer.

Egg mass

EGGS are in a soft, orange- to small grapefruit–sized mass. Individual eggs are black above and white below. Each egg has a wide layer of jelly, so the eggs are widely spaced in the mass. Many egg masses are usually laid on top of each other, and the tops of some masses are exposed above the water surface. The outer jelly layers may contain green algae. The mass usually spreads out and looks frothy just before hatching.

Hatchling

HATCHLINGS have a long, dark, streamlined look. The gills are long, like gnarly fingers. The tail is more than $1\frac{1}{2}$ times the body length. The dorsal fin is low, and the top edge angles up slightly from the back. It is almost opaque charcoal or black. (Also see Confusing Species Comparisons, p. 149.)

Tadpole

TADPOLES have a long, streamlined and dark appearance, because the tail is about twice the body length and the dorsal fin is low—it is the same height as the thickness of the tail trunk (at its base) or less. The mouth has three tooth rows on top and three to four on the

Tadpole

bottom. Tadpoles are charcoal to black, with fine, dull silver (or pale gold) flecks or speckling, especially on the belly, which looks silver or whitish. The dorsal fin is translucent dark gray, or densely speckled with black flecks. (Also see Confusing Species Comparisons, p. 150.)

JUVENILES have gold eyes that look out to the side. The dorsolateral folds are distinct on the full length of the back to the hip. They angle closer together on the lower back, which makes it look angular or flat. The snout is short, broad and slightly tapered. The vivid contrast between the dark mask and the light lip line, which is distinct to the end of the snout, make the snout appear pointed. The underside of the thigh is dull yellow or tan. (Also see Confusing Species Comparisons, p. 151.)

ADULTS have gold eyes that look to the side. The dorsolateral folds are distinct on the full length of the back to the hip. The hind legs are long. The webbing is stepped down on the inner edges of the toes. The back often has crisp-edged, usually round or angular, black spots, some with light centers. The groin has the same color pattern as further forward on the sides, or it may have a greenish wash. The underside of the thigh is translucent yellow-tan. The mask and lip line are usually distinct. The call is a quiet, deep, throaty chuckling. (Also see Confusing Species Comparisons, p. 152.)

Juvenile

Adult, showing groin

Top of left hind foot

SPOTTED FROG

Rana pretiosa BAIRD AND GIRARD, 1853

Adult

THE *Spotted Frog is quite variable in its color patterns and habits, and it has recently been split into at least two separate species. Spotted Frogs are widespread and common in many areas east of the Cascade Range and Coast Mountains, but they are now gone from many ponds in the western part of the range. These western ponds have either been filled in and covered by developments, or they have been taken over by the introduced Bullfrog, which both eats and outcompetes native amphibians. Also, Spotted Frogs no longer occur in ponds where bass and other warm-water, predatory fish were released. • Rana is Latin for 'frog'; pretiosa means 'of great value,' which may reflect a preference for eating this species as frog legs.*

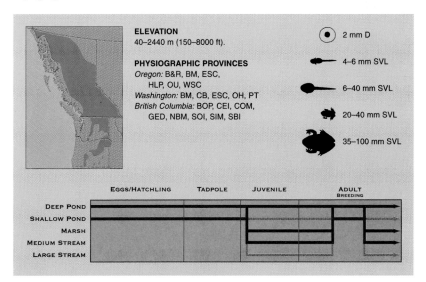

ELEVATION
40–2440 m (150–8000 ft).

PHYSIOGRAPHIC PROVINCES
Oregon: B&R, BM, ESC,
    HLP, OU, WSC
Washington: BM, CB, ESC, OH, PT
British Columbia: BOP, CEI, COM,
    GED, NBM, SOI, SIM, SBI

2 mm D

4–6 mm SVL

6–40 mm SVL

20–40 mm SVL

35–100 mm SVL

	EGGS/HATCHLING	TADPOLE	JUVENILE	ADULT BREEDING
DEEP POND				
SHALLOW POND				
MARSH				
MEDIUM STREAM				
LARGE STREAM				

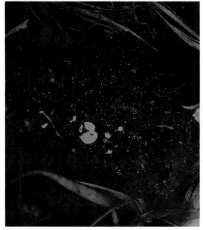
Egg mass

Hatchlings on top of egg mass

Tadpole

Tadpole

HABITATS: Spotted Frogs live in marshes, permanent ponds, lake edges and slow streams, usually where there is abundant aquatic vegetation. They breed in very shallow water—often a flooded meadow beside a pond or stream, or water pooled on top of flattened, dead vegetation at the edge of a pond—in early or mid spring. They lay their eggs on the bottom, usually on low vegetation. Hatchlings are often stranded in the water on top of the egg mass at first; then they swim in rain or snowmelt to deeper water. Tadpoles live in the warmest parts of ponds. Froglets and adults live in well-vegetated ponds, marshes or slow, weedy streams that meander through meadows.

EGGS are in a soft, orange-sized mass. Individual eggs are black above and white below. Each egg has a narrow layer of jelly, so fresh eggs are densely packed in the egg mass. Egg masses are laid on top of each other, sometimes in a pile of 20 or more. They are easily detached from the vegetation. Egg masses may float to the surface before hatching, where they spread out and look frothy.

HATCHLINGS have long gills that look like gnarly fingers. The huge tail is more than $1\frac{1}{2}$ times the body length. The translucent light gray dorsal fin is tall, and its top edge arches up from back near the base of the tail. The long tail and tall fin have a sweeping look, like a banner. (Also see Confusing Species Comparisons, p. 149.)

TADPOLES have a long tail (about twice the body length) and a tall, arching fin that looks like a huge banner. The dorsal fin is taller than the thickness of the tail trunk at its base (side view). The top edge of the dorsal fin arches up steeply from near the base of the tail. The mouth generally has two tooth rows on top and three on the bottom (but it may vary). Tadpoles are brown or gray, with dull gold flecks or speckling. The belly is pale gold. The dorsal fin is colorless or has scattered dark and light flecks. (Also see Confusing Species Comparisons, p. 150.)

JUVENILES have a slightly pointed snout, so the eyes may protrude beyond the sides of the snout (top view). The eyes are slightly upturned, and they may be dark, gold or yellow. The lip line is variable. The dorsolateral folds are inconspicuous, especially on the lower back, which looks plump and broad. (Also see Confusing Species Comparisons, p. 151.)

Juvenile

ADULTS have upturned eyes that are bright yellow (or chartreuse) or gold. The hind legs are short. The webbing extends along both sides of the toes almost to the tips. There are huge black spots on the back. These spots have blurred or sharply scalloped edges, and they often have light centers. The pattern in the groin is pale and often gray. The belly and underside of the thigh are opaque, with a mottling of brick red to orange-red or yellow-orange. The lip line is often long and distinct, and the mask is usually dull or absent. The call is a quiet, low-pitched, hollow knocking. (Also see Confusing Species Comparisons, p. 152.)

Adult

Underside of left hind foot

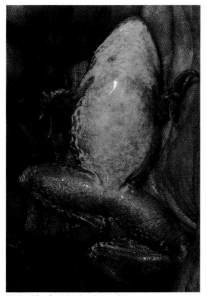

Underside of adult, showing groin

Adult

WOOD FROG

Rana sylvatica LeConte, 1825

Adult

THE *Wood Frog occurs in several very distinctive color patterns that may make you think there are several separate species. It is well adapted to a cold climate—it is the only North American amphibian that occurs north of the Arctic Circle—and it has a broad range across Canada. Most of the published information on the species in the Northwest has come from studies near Fairbanks, Alaska. Wood Frogs breed in the long daylight hours in spring, and they are remarkable for their vigorous gabbling calls and mating activity as the sun goes down and the water begins to freeze. Males have two throat pouches that balloon out on the upper back when they call, making the sound louder. • Rana is Latin for 'frog'; sylvatica means 'growing among trees,' which reflects the typical habitat for this species.*

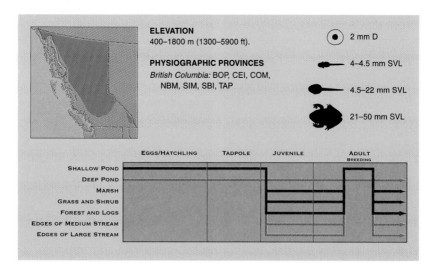

ELEVATION
400–1800 m (1300–5900 ft).

PHYSIOGRAPHIC PROVINCES
British Columbia: BOP, CEI, COM, NBM, SIM, SBI, TAP

2 mm D

4–4.5 mm SVL

4.5–22 mm SVL

21–50 mm SVL

	EGGS/HATCHLING	TADPOLE	JUVENILE	ADULT BREEDING
SHALLOW POND				
DEEP POND				
MARSH				
GRASS AND SHRUB				
FOREST AND LOGS				
EDGES OF MEDIUM STREAM				
EDGES OF LARGE STREAM				

HABITATS: Wood Frogs live in marshes, wet meadows, moist woodlands and brush. They breed in shallow pond edges, seasonal pools, flooded meadows, or the slow parts of streams in early spring to early summer, when ice is beginning to melt on the lakes. They attach their eggs around sedges or to submerged vegetation at or near the surface, in water 0.5 m deep or less. Many egg masses are usually laid together. Tadpoles live in the shallowest, warmest water available. Juveniles and adults live in wet meadows, riparian areas or moist brush or woods.

Egg mass

EGGS are in a soft egg mass that is plum- to small orange–sized (usually 5–7 cm in diameter). Individual eggs are black above and white below. Each egg has a narrow layer of jelly, so the eggs are tightly packed in the mass. Egg masses are laid close together or on top of each other, and the tops of some may be exposed above the water.

Hatchling

HATCHLINGS have long gills (almost as long as the head) that look like gnarly fingers. The head is small and narrower than the body. At the time of hatching, the tail is often shorter than the body, but it grows rapidly to slightly longer than the body. The top edge of the dorsal fin is low, and it angles slightly up from near the base of the tail.

Tadpole

TADPOLES have a very short, round body. The snout is short, and the large eyes are near the edge of the body (top view). The mouth has two to four tooth rows on top and three to four on the bottom. The tail is approximately 1½

Tadpole

times the body length or less. The tail trunk is dark along the top and light underneath. The dorsal fin is slightly taller than the thickness of the tail trunk at its base (side view). Tadpoles are black or very dark brown, with bright gold flecks that are most dense in several short lines that radiate out around the mouth (they look like golden whiskers). The belly is dark, with a silver sheen.

JUVENILES AND ADULTS have a long, pointed snout that is accentuated by the white lip line and the black mask, which are sharply distinct from the end of the snout to the shoulder. The eyes look outward. The dorsolateral folds are distinct to the hip. The hind legs are short (the lower leg is less than half of the snout-to-vent length). The webbing does not extend to the tips of the toes. Frogs occur in several distinct color types, one of which has a narrow, light stripe down the back. They may have irregular black markings on the back, particularly along the dorsolateral folds, or short, raised ridges outlined with black. The underside of the thigh is white. Males give a series of loud, throaty quacking calls (like gabbling geese).

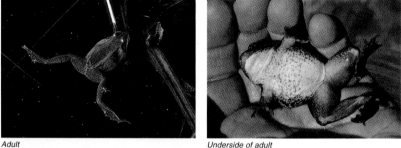

Adult   Underside of adult

Juvenile

NORTHERN LEOPARD FROG

Rana pipiens SCHREBER, 1782

Adult

NORTHERN *Leopard Frogs used to be caught in huge numbers for biology classes. The distinctive spots on its back make it a gorgeous sight, but it may be disappearing from this region and throughout its range. Studies indicate that some entire populations have been killed off by diseases that may be related to environmental stresses. We were lucky enough to find the only Northern Leopard Frog egg mass seen in Oregon or Washington for quite a few years. We made several visits one spring to a pond with many Northern Leopard Frogs and potentially excellent habitat for their eggs and tadpoles. When the water level rose, however, introduced carp entered the pond, and they ate all the vegetation and left a denuded site with no habitat for frog eggs or tadpoles—and no frogs. • Rana is Latin for 'frog'; pipiens means 'peeping,' which hardly does justice to the male's astonishing repertoire of rubbery putterings and trills—it may instead refer to the squeak it makes when it is caught or chased into the water.*

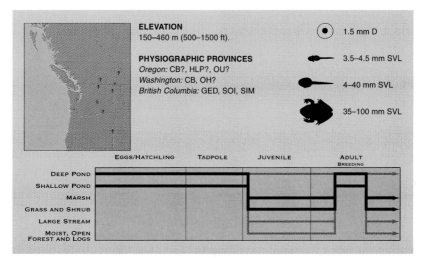

ELEVATION
150–460 m (500–1500 ft).

PHYSIOGRAPHIC PROVINCES
Oregon: CB?, HLP?, OU?
Washington: CB, OH?
British Columbia: GED, SOI, SIM

1.5 mm D

3.5–4.5 mm SVL

4–40 mm SVL

35–100 mm SVL

	EGGS/HATCHLING	TADPOLE	JUVENILE	ADULT BREEDING
DEEP POND				
SHALLOW POND				
MARSH				
GRASS AND SHRUB				
LARGE STREAM				
MOIST, OPEN FOREST AND LOGS				

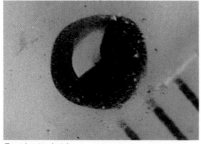
Egg about to hatch

Hatchling

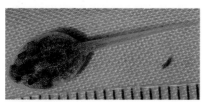
Small tadpole

Large tadpole

HABITATS: Northern Leopard Frogs live in marshes, wet meadows, riparian areas and moist, open woods. They breed in ponds or lake edges with fairly dense aquatic and emergent vegetation in mid spring, and they attach their eggs to submerged vegetation well below the surface, in water 0.5 m deep or more. Hatchlings cling to the egg mass or nearby vegetation. Tadpoles live in dense aquatic vegetation. Juveniles and adults live in aquatic vegetation in ponds, and in adjacent grass, sedge, weeds or brush.

EGGS are in a grapefruit-sized mass (usually 8–13 cm in diameter). Individual eggs are small, and they have a thin jelly layer, so they are densely packed.

HATCHLINGS have gills that are distinctly shorter than the head length. The tail is barely longer than the body. The dorsal fin is low, and its top edge slopes only slightly up from near the base of the tail (side view).

TADPOLES have a tail that is not much longer than the body until the tadpole is larger than approximately 25 mm SVL. Tadpoles often get quite large just before metamorphosis, and the tail grows to more than $1\frac{1}{2}$ times the body length. The dorsal fin is up to the same height as the thickness of the tail trunk at its base (side view). The top edge of the dorsal fin slopes slightly up from near the base of the tail. The mouth has two tooth rows on top and three on the bottom. Tadpoles are dark brown or gray, with an increasingly dense covering of silver or pale gold blotches on the underside (smaller on the back). The belly may be transparent, making the guts visible. There may be one or two whitish lines near the mouth. The tail is pale tan, and the dorsal fin is colorless (or it may have fine, dark flecks).

JUVENILES AND ADULTS have a long, narrow, straight-sided snout. The eyes protrude slightly beyond the sides of the snout (top view). The dorsolateral folds, which extend the full length of the back, are conspicuous. They are white or cream, in sharp contrast to the color of the back. The legs are long (the lower leg is more than half of the snout-to-vent length). The webbing does not extend to the tips of the toes. Frogs are bright green or tan, with very large, unmistakable brown or black blotches that are round or oval and have light borders. Juveniles may lack these spots. The underside, including the thighs, is opaque white. The calls are loud and varied, rubbery chuckles and gabbles.

Juvenile

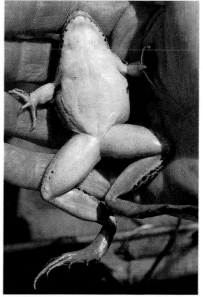

Underside of adult

FOOTHILL YELLOW-LEGGED FROG

Rana boylii BAIRD, 1854

Adult

THE long common name of this species is needed to distinguish it from the related Mountain Yellow-legged Frog of California. Most true frogs live in ponds, or at least lay their eggs in still water, but the Foothill Yellow-legged Frog lives in streams or rivers. Although it is a member of the family Ranidae, this frog has some similarities to the Tailed Frog, which also lives in streams. The tadpoles of both species have flattened shapes and enlarged mouths to cling to rocks and keep from being swept away. The tadpoles forage by scraping algae and diatoms off the rocks. Even the adults of these two species are similar—they have grainy skin and a vague, mottled color pattern, both of which are excellent camouflage in rocks.

• Rana is Latin for 'frog'; boylii honors Dr. C.C. Boyle, who collected the first series of this frog.

ELEVATION
Sea level to 550 m (1800 ft).

PHYSIOGRAPHIC PROVINCES
Oregon: KM, OCR, WSC, WIV

- 2 mm D
- 3–4 mm SVL
- 4–20 mm SVL
- 16–75 mm SVL

	EGGS/HATCHLING	TADPOLE	JUVENILE	ADULT BREEDING
MEDIUM STREAM				
LARGE STREAM				
SMALL STREAM				

HABITATS: Foothill Yellow-legged Frogs live in sections of low-gradient streams with exposed bedrock or rock and gravel substrates. They lay their eggs in late spring or early summer, and they attach them to the bottom of quiet scour-pools or riffles in gentle-gradient streams, often where there is only slight flow from the main river. Hatchlings cling to the egg mass initially, and then to rocks. Tadpoles live in pools that often have a connection to the main river flow, but little or no silt. Froglets live in pools with gravel and cobbles. Adults live in pool edges (often in a deep pool with sedge clumps around the edge), in bedrock at the edge of the main channel, or under cobbles at the bottom of the pool.

Egg mass

EGGS are in a soft, orange- to grapefruit-sized mass. Individual eggs are black above and white below. The jelly layer around each egg is thin, so the eggs are densely packed in the cluster.

HATCHLINGS have a small mouth that is oriented down, and they cling by suction to rocks in gently flowing water. The tail is short, and the back half of the tail trunk is colorless. The dorsal fin is low.

Hatchling

TADPOLES are long and low, and the underside looks flattened (side view). They can cling by suction to rocks, because the mouth is large and oriented down. There are six to seven tooth rows on top and five to six on the bottom, but they are difficult to see. The top edge of the dorsal fin is low at the base of the tail, but it flares higher halfway along the tail. Tadpoles are dark gray, with tiny, gold and whitish flecks all over. The tip of the tail trunk is colorless. The tail fins are colorless, or they have a few dark flecks (denser near the base of the tail).

Tadpole

Underside of tadpole

JUVENILES AND ADULTS have indistinct dorsolateral folds (or none). The hind legs are long, and the webbing is full. Frogs are gray or brown, speckled with darker dots, and they may have some reddish patches. The throat often has dark gray mottling. The underside is cream or white, with light or bright yellow on the sides of the belly and the underside of the thighs. The skin is grainy, and in color and texture it looks like a rock's surface. The call is a quiet, throaty, short, low-pitched trill.

Juvenile

Underside of adult

BULLFROG (introduced)

Rana catesbeiana SHAW, 1802

Adult

A LTHOUGH *the Bullfrog is a wonderful part of many ecosystems in the eastern United States, it is the bad guy of the Pacific Northwest. It was introduced here in the 1920s or 1930s to be raised for food, because it has big, meaty legs. It is much larger than our native frogs, and through no fault of its own, it has eaten, outcompeted and may have displaced native amphibians at warm ponds in many areas within this region. There is ample evidence that Bullfrogs also eat turtle hatchlings, ducklings and other birds, and they have few predators in this region. Now there are even alarming anecdotes that it has invaded some higher-elevation ponds that were thought to be too cold for it. Disturbance and development around wetlands often create warmer conditions. Are wetland restoration projects just Bullfrog nurseries? • Rana is Latin for 'frog'; catesbe·iana means 'belonging to Catesby,' and it honors an English naturalist who studied the natural history of Florida and North Carolina in the early 1700s.*

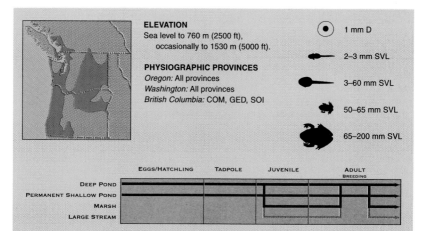

Egg mass

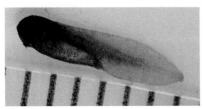
Hatchling

Yearling tadpole

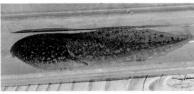
Two-year-old tadpole

HABITATS: Bullfrogs live in very warm and sunny permanent ponds, marshes or slow river backwaters, and they may forage in streams and temporary ponds. They lay their eggs in summer on the warm water surface. Tadpoles live in very warm shallows and dense aquatic vegetation. Froglets and adults live in permanent water among aquatic weeds, or in vegetation on the banks of ponds or slow rivers.

EGGS are laid in a broad, often frothy sheet of jelly, so they look like poppy seeds scattered on a patch of slime. They may become covered with algae. The egg mass is laid on the water surface, but it slumps onto submerged or aquatic vegetation just before hatching. Individual eggs are black above and white below. They are very small, and they have a wide layer of jelly.

HATCHLINGS are suprisingly tiny, with a very slender body and short tail. The gills are short, and they may already be covered before hatching. Hatchlings are pale gray-tan, and the yolk is easily visible in the belly.

TADPOLES have a long body, a short tail and prominent nostrils when they are small. After a tadpole has grown to about 15 mm SVL, its body is long and arrowhead-shaped, with a slightly pointed snout. The eyes are orange or bronze. The mouth has two to three tooth rows on top and three on the bottom. The tail trunk is massive; the dorsal fin is not as tall as the thickness of the trunk at its base, and it begins behind the body (side view). Small tadpoles (3–12 mm SVL) are dark gray or brown, with gold flecks in three bands across the body. Mid-sized tadpoles (about 12–25 mm SVL) are green or olive-tan, with sharp-edged, black polka-dots and gold patches. The largest tadpoles have dark gray mottling on the sides, opaque yellow undersides and less distinct black polka-dots. In this region, tadploes usually metamorphose after two years.

Adult male

JUVENILES are as large as the adults of some other species. The conspicuous, large ear drum is partially surrounded by a ridge or groove. There are no dorsolateral folds. Juveniles are green to brown, with sharp-edged, black polka-dots on the back and on the top of the snout. The eyes are orange or bronze.

Juvenile with tail remnant

ADULTS look like juveniles (only bigger). They are green, tan or dark brown. The underside is whitish, with large, gray mottlings on many larger individuals. The ear drum is generally the same color as the body, or it may have dark speckling around the edge. The upper lip is bright green, with blurred edges, and the back and top of the snout may have gray or vague, dark spots. The eyes are gold. Males have a yellow throat, and the eardrum is larger than the eye. The call is a loud, deep-pitched, two-part droning. Juveniles and adults squeak when they are scared into the water.

GREEN FROG (introduced)

Rana clamitans LATREILLE, 1801

Adult

THE *Green Frog is a handsome frog that is valuable in its natural range in the eastern United States, and so far it is not considered to have caused problems for our native wildlife or to have spread from the small areas where it was introduced. Only a few small, isolated populations are known. Although the Green Frog is smaller and milder than the Bullfrog, it should be watched closely, and further introductions should be strongly discouraged. Both Green Frogs and Bullfrogs are hard to catch because they leap for the pond at the first approach. When they stare at a bright flashlight at night, however, they seem unable to see a hand reaching out to grab them.* • Rana *is Latin for 'frog';* clamitans *means 'complaining noisily,' which reflects this species's loud, continuously repeated, twanging calls.*

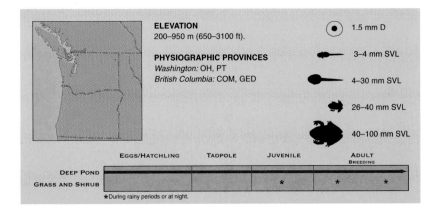

ELEVATION
200–950 m (650–3100 ft).

PHYSIOGRAPHIC PROVINCES
Washington: OH, PT
British Columbia: COM, GED

1.5 mm D
3–4 mm SVL
4–30 mm SVL
26–40 mm SVL
40–100 mm SVL

	EGGS/HATCHLING	TADPOLE	JUVENILE	ADULT BREEDING
DEEP POND				
GRASS AND SHRUB			*	* *

*During rainy periods or at night.

HABITATS: Green Frogs live in permanent lakes, ponds or deep ditches with
dense aquatic vegetation. They lay their
eggs in early to late summer on the water
surface, spread out over aquatic vegetation. Tadpoles live in aquatic vegetation. Juveniles and adults also live in
aquatic vegetation, as well as in emergent plants or dense vegetation along the
bank.

Hatchling

EGGS are very similar to Bullfrog eggs,
but the egg mass is smaller (usually less
than 30 cm in diameter). The eggs are in
a broad, frothy sheet of jelly, so they look
like poppy seeds scattered on a patch of

Tadpole

slime. The egg mass is laid on the water surface, but it slumps onto submerged vegetation just before hatching. Individual eggs are black above and white below. They
are very small, and they have a wide layer of jelly.

HATCHLINGS are very similar to Bullfrog hatchlings. They are tiny, pale gray-tan,
and the yolk is easily visible in the belly. The gills are short, and they are often
covered before hatching. The tail is short.

TADPOLES are very similar to Bullfrog tadpoles when they are small (4–10 mm
SVL). They have conspicuous nostrils, but the bands of gold flecks across the body
are not as distinct. Larger tadpoles have a long, oval body (widest at the spiracle)
and a pointed snout (top view). There are two tooth rows on top and three on the
bottom. The tail trunk is massive; the dorsal fin is not as tall as the thickness of the
trunk at its base, and it begins behind the body. Large tadpoles are olive green, with
vague, black, blurred-edge polka-dots that may be obscured by dense gold patches on
the back. The belly is cream, with a mottled edge between the dorsal and the ventral
colors. The mouth and surrounding area are dusky gray. The eyes are pale bronze.

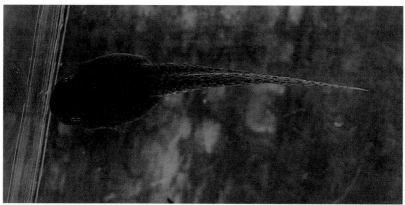
Tadpole

Juvenile with tail remnant

Underside of adult

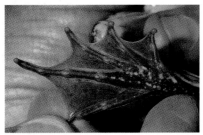

Underside of right hind foot

JUVENILES have a ridge or groove partially surrounding the eardrum. The dorsolateral folds may not be evident until a froglet is about 40 mm SVL. Juveniles are mottled gray, tan or dull gold, grading to whitish on the underside. The upper lip is mottled green, and a broken, light green lip line extends to the shoulder.

ADULTS have a large, conspicuous eardrum (especially on the male) that has a lighter center like a bull's-eye, and it is partially surrounded by a ridge or groove. The dorsolateral folds are distinct, but they only occur on the upper part of the back. The webbing on the hind feet does not extend to the toe tips, and it is deeply scalloped. Adults are green or brown, with a bright green upper lip and sharp-edged lip line that extends to the shoulder. Black polka-dots may be present just on top of the snout, or they may boldly but irregularly cover the head, back and sides. The underside is whitish, but it may have some dark mottling on the chest. Males have a bright yellow throat. The call is a loud, short, guttural twang, like a plucked banjo string.

Juvenile

IDENTIFYING AMPHIBIANS

Western Toad

IDENTIFICATION KEYS

Each key is arranged as a series of either/or choices. **Always** start at the beginning of each key, and read both options. To move to the next step in the key, follow the instructions at the end of the option that best describes your animal. Continue to select the better of each pair of choices until you have arrived at a species name.

Don't give up if your amphibian doesn't match the description precisely; amphibian species can be widely variable in many characteristics. If you are unsure about choosing between two options, explore each option to see which fits best. If you get to the end of the key, and the resulting identification just does not make sense, go back to the beginning of the key and try the other choices where you were not sure.

Sometimes you will find an amphibian with several features indicating one species and others indicating another species. Give greater weight to the features listed first in the keys, and decide which option shows the majority of characteristics. Remember that this is an artificial key that uses only features that can be determined in the field without damage to the animal.

Keys to the Eggs of Pond-breeding Amphibians

These keys include salamander and frog eggs that are laid in the open, in still water or in slow sections of streams. The eggs of terrestrial salamanders and stream-breeding amphibians are not included because those species hide their eggs, and we do not recommend searching for them. The descriptions are of eggs in the early development stages. In later stages, you may be able to use egg mass characteristics or the key to hatchlings to identify the species. Egg mass measurements include all the jelly layers. Individual egg measurements refer to the diameter of the egg (ovum) and do not include the outer jelly layers.

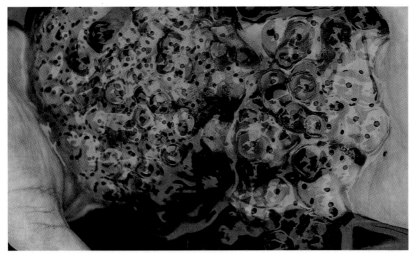

Eggs of the Spotted Frog (left) and the Red-legged Frog (right)

These keys contain general descriptions, and they are presented with the following precautions:

- occasionally, eggs get mislaid singly instead of in a cluster
- colors may vary considerably
- water levels may change after the eggs are laid
- eggs may become coated with silt, making them difficult to recognize
- invertebrates often lay their eggs in the same environments as amphibians; their eggs are generally smaller individually (less than 0.5 mm), they are sometimes pale in color, and they are often in regular rows within the mass
- jelly blobs with no internal structure are often algae.

There are separate keys for the western and eastern portions of the region. However, the east slopes of the Cascade Range and Coast Mountains may include species from both sides of the mountains, so you may have to try both keys.

The western portion includes the following physiographic provinces:

- *Oregon:* East Slope Cascades (western portion), Klamath Mountains, Oregon Coast Range, West Slope Cascades, Western Interior Valleys
- *Washington:* Eastern Slopes, Washington Cascades (western portion); Olympic Peninsula & Southwestern Washington; Puget Trough; Western Slopes & Crest, Washington Cascades
- *British Columbia:* Coast & Mountains, Georgia Depression.

The eastern portion includes the following physiographic provinces:

- *Oregon:* Basin & Range, Blue Mountains, Columbia Basin, East Slope Cascades (eastern portion), High Lava Plains, Owyhee Uplands
- *Washington:* Blue Mountains; Columbia Basin; Eastern Slopes, Washington Cascades (eastern portion); Okanogan Highlands
- *British Columbia:* Boreal Plains, Central Interior, Northern Boreal Mountains, Southern Interior, Southern Interior Mountains, Sub-boreal Interior, Taiga Plains.

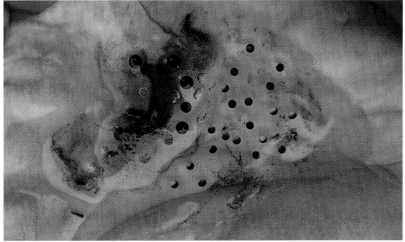

Eggs of the Long-toed Salamander (left) and the Pacific Treefrog (right)

KEY TO THE EGGS OF WESTERN OREGON, WASHINGTON AND BRITISH COLUMBIA

1

1a • Eggs are single-file in long jelly strings, which may be wound around and cover an extensive area.
.... **Western Toad**

1b • Eggs are **either** in a round jelly cluster **or** single **or** in a broad sheet of jelly.
.... **2**

2

2a • Eggs are in a broad sheet of jelly.
• Egg mass is frothy, is laid on water surface above dense aquatic or emergent vegetation.
• Egg mass later loses frothiness and slumps onto underlying vegetation.
.... **3**

2b • Eggs are **either** in a round jelly cluster **or** single.
• Eggs are laid near or below water surface.
• When water levels change, or just before hatching, egg mass may float to surface and spread out, and it may look frothy.
.... **4**

3

3a • Egg mass is usually less than 30 cm in diameter.
.... **Green Frog**

3b • Egg mass is usually more than 30 cm in diameter.
.... **Bullfrog**

4

4a • Egg mass is a firm jelly ball that is smooth or slightly lumpy with additional jelly layers around entire mass.
• Egg mass is attached around submerged twig.
• Green algae usually grows in each egg's jelly layer.
• Eggs are tan or gray-tan above, cream below.
.... **Northwestern Salamander**

4b • Eggs are **either** single **or** in a soft, lumpy cluster.
• Egg mass may or may not be attached.
• Green algae may grow in outer jelly layers of mass just before hatching.
• Eggs are various colors.
.... **5**

5

5a • Eggs are single.
.... **6**

5b • Eggs are in a cluster.
.... **7**

6

6a • Layer of jelly around egg is thinner than egg diameter.
• Egg is attached in vegetation, usually well hidden (tucked between stems or with leaf wrapped across egg).
• Egg is tan above, cream below.
.... **Roughskin Newt**

6b • Layer of jelly around egg is wider than egg diameter.
• Egg is attached or rests on bottom, usually not hidden.
• Egg is black or dark brown above, white below.
.... **Long-toed Salamander**

7

7a • Egg cluster is small, usually less than 5 cm in diameter (small grape– to small plum–sized).
.... **8**

7b • Egg cluster is large, usually more than 5 cm in diameter (large plum– to cantaloupe-sized).
.... **9**

8

8a • Eggs are small (1.5 mm D or less).
• Eggs are packed closely together (because jelly layer is thin).
• Eggs are tan to gray-brown above, yellow-gold to cream below.
.... **Pacific Treefrog**

8b • Eggs are large (2 mm D or more).
• Eggs are widely spaced (because jelly layer is wide).
• Eggs are black or dark brown above, white below.
.... **Long-toed Salamander**

9

9a • Egg mass is in gently flowing water, slight riffle or river-scoured pool.
• Egg mass is attached to rock.
.... **Foothill Yellow-legged Frog**

9b • Egg mass is in still water (may have slight flow).
• Egg mass may be attached to vegetation.
.... **10**

10

10a • Eggs are very large (3 mm D).
• Eggs are usually laid below water surface in at least 50 cm of water.
• Egg masses are laid separately (but may be close together).
• Egg mass is loosely attached around submerged stem or branch, or is stuck to bottom vegetation.
.... **Red-legged Frog**

10b • Eggs are large (2 mm D).
• Eggs are usually at water surface in less than 50 cm of water (top of mass is often above surface).
• Egg masses are laid on top of each other.
• Egg mass rests on or is stuck to low or flattened vegetation.
.... **11**

11

11a • Eggs are packed close together (because jelly layer is thin).
.... **Spotted Frog**

11b • Eggs are widely spaced (because jelly layer is wide).
.... **Cascades Frog**

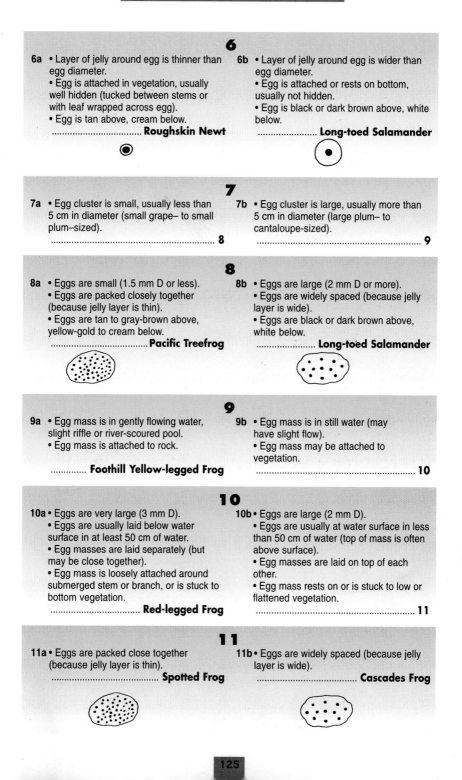

KEY TO THE EGGS OF EASTERN OREGON, WASHINGTON AND BRITISH COLUMBIA

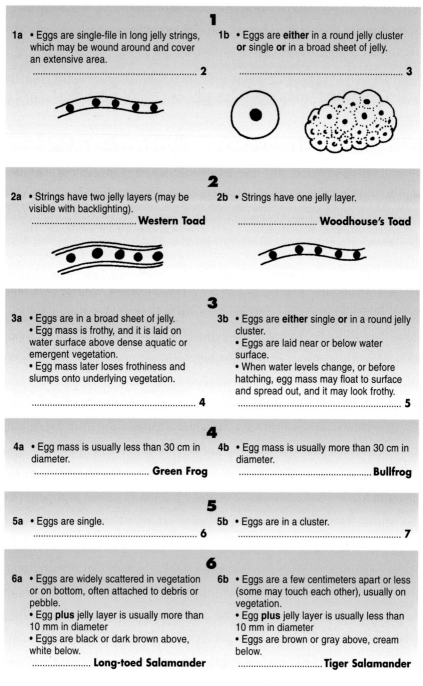

1

1a • Eggs are single-file in long jelly strings, which may be wound around and cover an extensive area.
.... **2**

1b • Eggs are **either** in a round jelly cluster **or** single **or** in a broad sheet of jelly.
.... **3**

2

2a • Strings have two jelly layers (may be visible with backlighting).
.... **Western Toad**

2b • Strings have one jelly layer.
.... **Woodhouse's Toad**

3

3a • Eggs are in a broad sheet of jelly.
• Egg mass is frothy, and it is laid on water surface above dense aquatic or emergent vegetation.
• Egg mass later loses frothiness and slumps onto underlying vegetation.
.... **4**

3b • Eggs are **either** single **or** in a round jelly cluster.
• Eggs are laid near or below water surface.
• When water levels change, or before hatching, egg mass may float to surface and spread out, and it may look frothy.
.... **5**

4

4a • Egg mass is usually less than 30 cm in diameter.
.... **Green Frog**

4b • Egg mass is usually more than 30 cm in diameter.
.... **Bullfrog**

5

5a • Eggs are single.
.... **6**

5b • Eggs are in a cluster.
.... **7**

6

6a • Eggs are widely scattered in vegetation or on bottom, often attached to debris or pebble.
• Egg **plus** jelly layer is usually more than 10 mm in diameter
• Eggs are black or dark brown above, white below.
.... **Long-toed Salamander**

6b • Eggs are a few centimeters apart or less (some may touch each other), usually on vegetation.
• Egg **plus** jelly layer is usually less than 10 mm in diameter
• Eggs are brown or gray above, cream below.
.... **Tiger Salamander**

7

7a • Egg cluster is small, usually less than 5 cm in diameter (small grape– to small plum–sized).
.... **8**

7b • Egg cluster is large, usually more than 5 cm in diameter (large plum– to grapefruit-sized).
.... **11**

8

8a • Eggs are in a randomly shaped cluster and one egg can be detached easily.
• Eggs are small (1 mm D or more).
• Eggs are gray above, cream below.

.... **Great Basin Spadefoot**

8b • Egg mass is rounded and holds together (difficult to detach one egg).
• Eggs are **either** small (1 mm D or less) **or** large (2 mm D or more).
• Eggs are tan, brown or black above, white, cream or yellow-gold below.
.... **9**

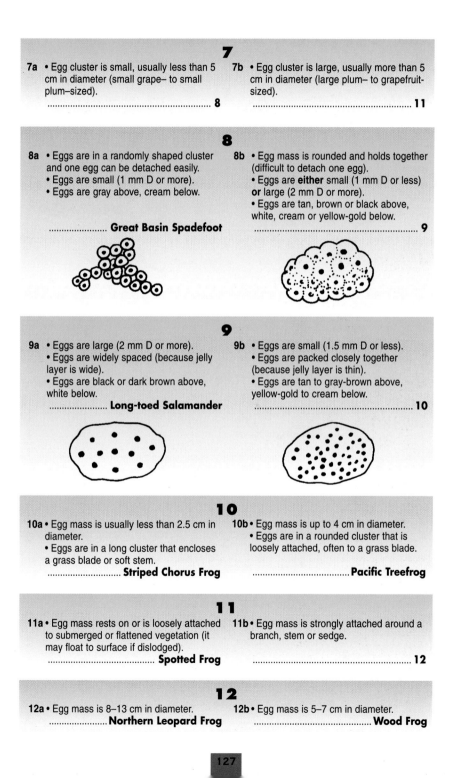

9

9a • Eggs are large (2 mm D or more).
• Eggs are widely spaced (because jelly layer is wide).
• Eggs are black or dark brown above, white below.
.... **Long-toed Salamander**

9b • Eggs are small (1.5 mm D or less).
• Eggs are packed closely together (because jelly layer is thin).
• Eggs are tan to gray-brown above, yellow-gold to cream below.
.... **10**

10

10a • Egg mass is usually less than 2.5 cm in diameter.
• Eggs are in a long cluster that encloses a grass blade or soft stem.
.... **Striped Chorus Frog**

10b • Egg mass is up to 4 cm in diameter.
• Eggs are in a rounded cluster that is loosely attached, often to a grass blade.
.... **Pacific Treefrog**

11

11a • Egg mass rests on or is loosely attached to submerged or flattened vegetation (it may float to surface if dislodged).
.... **Spotted Frog**

11b • Egg mass is strongly attached around a branch, stem or sedge.
.... **12**

12

12a • Egg mass is 8–13 cm in diameter.
.... **Northern Leopard Frog**

12b • Egg mass is 5–7 cm in diameter.
.... **Wood Frog**

KEY TO HATCHLING SALAMANDERS

Only pond-breeding salamanders have a recognizable hatchling stage.
See the glossary of color and markings (p.171).

1

1a • Gill stalks are thick and longer than head length.
• Balancers are absent.
• Lower pair of gills often point down below body (front view), and tips may act as balancers.
• Gills are swept back along body (top view).
.... **Tiger Salamander**

1b • Gill stalks are thin and shorter than head length.
• Balancers are present for first few days.
• Lower pair of gills point out to sides of body (front view).
• Gills curve out, with tips bent away from body (top view).
.... **2**

2

2a • Snout is narrow, and eyes look forward, because front corners are closer together than back corners (top view).
• Front foot is visible beyond gills, because upper part of leg is longer than lower part (top view).
• May have at least one row of indistinct yellow dots along side, and gills are speckled.
.... **Roughskin Newt***

2b • Snout is broad, and eyes look out to sides (top view).
• Front foot is usually hidden beneath gills, because upper part of leg is shorter than lower part (top view).
• No yellow dots or speckles.
.... **3**

3

3a • Head is short (about $^1/_5$ of TL), only slightly wider than body (top view).
• Gill stalk has few side filaments that occur along full length of stalk.
• Hind legs usually appear at 25 mm TL or more.

.... **Northwestern Salamander***

3b • Head is large (about $^1/_3$ of TL), distinctly wider than body (top view).
• Gill stalk has few side filaments that occur only near base of stalk, leaving a long spike.
• Hind legs usually appear at 25 mm TL or less.
.... **Long-Toed Salamander***

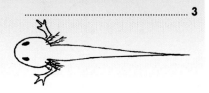

* Also see Confusing Species Comparisons (p. 146).

KEY TO LARVAL SALAMANDERS

This key can be used to identify both immature and neotenic larvae of aquatic salamanders. See the glossary of color and markings (p. 171).

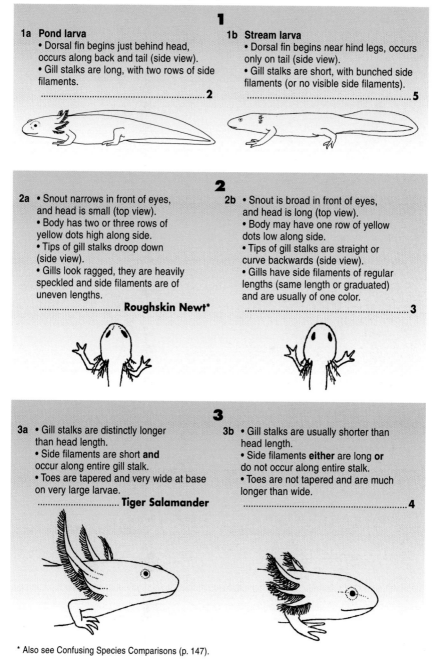

1

1a Pond larva
  • Dorsal fin begins just behind head, occurs along back and tail (side view).
  • Gill stalks are long, with two rows of side filaments.
.... 2

1b Stream larva
  • Dorsal fin begins near hind legs, occurs only on tail (side view).
  • Gill stalks are short, with bunched side filaments (or no visible side filaments).
.... 5

2

2a • Snout narrows in front of eyes, and head is small (top view).
  • Body has two or three rows of yellow dots high along side.
  • Tips of gill stalks droop down (side view).
  • Gills look ragged, they are heavily speckled and side filaments are of uneven lengths. .... **Roughskin Newt***

2b • Snout is broad in front of eyes, and head is long (top view).
  • Body may have one row of yellow dots low along side.
  • Tips of gill stalks are straight or curve backwards (side view).
  • Gills have side filaments of regular lengths (same length or graduated) and are usually of one color.
.... 3

3

3a • Gill stalks are distinctly longer than head length.
  • Side filaments are short **and** occur along entire gill stalk.
  • Toes are tapered and very wide at base on very large larvae.
.... **Tiger Salamander**

3b • Gill stalks are usually shorter than head length.
  • Side filaments **either** are long **or** do not occur along entire stalk.
  • Toes are not tapered and are much longer than wide.
.... 4

* Also see Confusing Species Comparisons (p. 147).

4

4a • Poison glands are visible on large larvae, at parotoid area and on tail ridge below fin.
• Gills appear full and like ostrich plumes (side filaments are long and willowy).
• Side filaments occur along full length of stalk.
• Larvae are often longer than 35 mm SVL.
.... **Northwestern Salamander***

4b • Poison glands are not apparent.
• Gills appear organized, with side filaments of graduated lengths (shorter near tip).
• Side filaments occur near base of stalk, leaving a long spike at tip.
• Larvae are rarely longer than 35 mm SVL.
.... **Long-toed Salamander***

5

5b • Underside is distinctly yellow.
• Gill stalks are tiny, with few or no visible side filaments.
• White flecks on back and sides.
.... **Torrent Salamanders**

5b • Underside is white or gray.
• Gill stalks are short, with bunched side filaments.
• No flecks.
....**6**

6

6a • Body is long (toes do **not** touch if legs are adpressed).
• Gill filaments are shorter than gill stalks, so gills appear separate, like shaving brushes.
• Head appears almost as narrow as body (top view).
• Head is usually rectangular, about as narrow at gills as at eyes (top view).
.... **Cope's Giant Salamander***

6b • Body is short (toes touch or overlap if legs are adpressed).
• Gill filaments are longer than gill stalks and often cover them, so gills appear as one bush on each side of head.
• Head is massive, appears much wider than body (top view).
• Head is usually wider at gills than at eyes (top view).
.... **Pacific Giant Salamander***

* Also see Confusing Species Comparisons (pp. 147 & 148).

KEY TO METAMORPHOSED AND TERRESTRIAL SALAMANDERS

See the glossary of color and markings (p. 171).

1

1a • Skin is grainy, dry and dull (less so when in water).
• Underside is orange and dorsal color is brown.
• Costal grooves are absent.
.... **Roughskin Newt**

1b • Skin is smooth and shiny.
• Color is various.
• Costal grooves are present, but may be indistinct.
....2

2

2a • Color is solid brown to brown-gray.
• Poison glands form conspicuous clusters of pale dots at parotoid areas, on back and on tail ridge (barely visible on some juveniles).
• Costal grooves are pronounced.
.... **Northwestern Salamander**

2b • Color is various, often with contrasting colors, a bold pattern, a dorsal stripe or metallic dots.
• Poison glands are not noticeable.
• Costal grooves are present, but they may be indistinct.
....3

3

3a • Color is translucent honey-tan (or olive); underside is distinctly yellow.
• Eyes are huge, comically perched near end of short snout.
.... **Torrent Salamanders**

3b • Color is various, often with a bold pattern, a dorsal stripe or metallic dots.
• Eyes are otherwise.
....4

4

4a • Fourth toe on hind foot is noticeably longer than other toes.
.... **Long-toed Salamander**

4b • Fourth toe on hind foot is about the same length as second and third toes.
....5

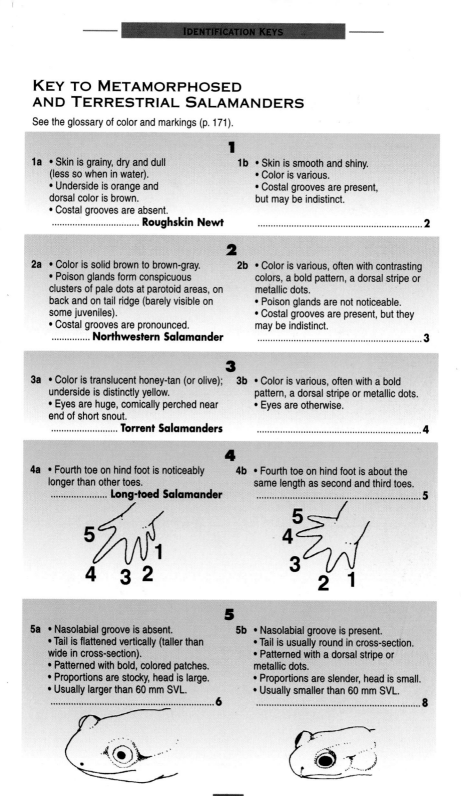

5

5a • Nasolabial groove is absent.
• Tail is flattened vertically (taller than wide in cross-section).
• Patterned with bold, colored patches.
• Proportions are stocky, head is large.
• Usually larger than 60 mm SVL.
....6

5b • Nasolabial groove is present.
• Tail is usually round in cross-section.
• Patterned with a dorsal stripe or metallic dots.
• Proportions are slender, head is small.
• Usually smaller than 60 mm SVL.
....8

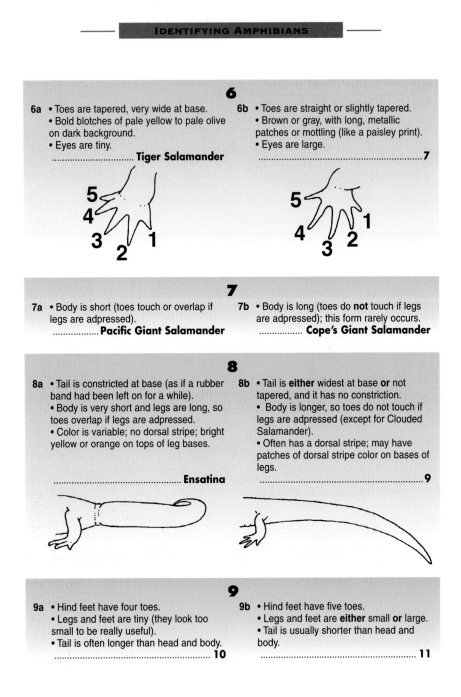

6

6a • Toes are tapered, very wide at base.
• Bold blotches of pale yellow to pale olive on dark background.
• Eyes are tiny.
.... **Tiger Salamander**

6b • Toes are straight or slightly tapered.
• Brown or gray, with long, metallic patches or mottling (like a paisley print).
• Eyes are large.
....7

7

7a • Body is short (toes touch or overlap if legs are adpressed).
.... **Pacific Giant Salamander**

7b • Body is long (toes do **not** touch if legs are adpressed); this form rarely occurs.
.... **Cope's Giant Salamander**

8

8a • Tail is constricted at base (as if a rubber band had been left on for a while).
• Body is very short and legs are long, so toes overlap if legs are adpressed.
• Color is variable; no dorsal stripe; bright yellow or orange on tops of leg bases.

.... **Ensatina**

8b • Tail is **either** widest at base **or** not tapered, and it has no constriction.
• Body is longer, so toes do not touch if legs are adpressed (except for Clouded Salamander).
• Often has a dorsal stripe; may have patches of dorsal stripe color on bases of legs.
.... 9

9

9a • Hind feet have four toes.
• Legs and feet are tiny (they look too small to be really useful).
• Tail is often longer than head and body.
.... 10

9b • Hind feet have five toes.
• Legs and feet are **either** small **or** large.
• Tail is usually shorter than head and body.
.... 11

10

10a • Toes are round or oblong, so feet look cat-like.
  • Body is dark, with a dorsal stripe consisting of small patches of pink to brick red organized in a chevron design.
  • Underside is dark, with tiny white dots.
.... **California Slender Salamander**

10b • Toes have slightly widened, round tips.
  • Body is dark, with a dorsal stripe consisting of unorganized patches of brick red to orange and gold.
  • Underside is dark, with bold white patches.
.... **Oregon Slender Salamander**

11

11a • Toes are widest near tip.
  • Outer (fifth) toe of hind foot is ³/₄ the length of fourth toe or longer.
  • Legs, feet and toes are large and strong.

.... **12**

11b • Toes are straight **or** widest near base.
  • Outer (fifth) toe of hind foot is ¹/₂ the length of fourth toe or shorter.
  • Legs, feet and toes are small and delicate.

.... **13**

12

12a • Toes touch or just overlap if legs are adpressed.
  • Tips of toes are wide and square.
  • Outer (fifth) toe of hind foot is same length as fourth toe.
  • Underside is gray.
.... **Clouded Salamander**

12b • Toes do not touch if legs are adpressed.
  • Tips of toes are wide and rounded.
  • Outer (fifth) toe of hind foot is ³/₄ the length of fourth toe.
  • Underside is black with white flecks.
.... **Black Salamander**

13

13a • Parotoid glands are present, but they may be hard to see.
  • Body appears short, usually has 14 costal grooves in total (or less than three intercostal folds between toes of adpressed legs).
Van Dyke's Salamander
.... and **Coeur d'Alene Salamander**

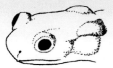

13b • Parotoid glands are absent.
  • Body appears long, usually has 15 or more costal grooves in total (or usually more than three intercostal folds between toes of adpressed legs).

.... **14**

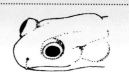

14

14a • Outer (fifth) toe of hind foot is just a nubbin ($^1/_4$ or less the length of fourth toe).
• Adult has salmon pink underside.
• Juvenile has dark underside with yellow or pink patch on belly.
.... **Larch Mountain Salamander**

14b • Outer (fifth) toe of hind foot is $^1/_3$ to $^1/_2$ the length of fourth toe.
• Underside is gray, tan or black.
.... **15**

15

15a • Dorsal stripe is mustard or olive tan, with ragged edges, and does not extend to tip of tail.
• Pieces of dorsal stripe color are also on sides of body and tops of entire legs.

.... **Dunn's Salamander**

15b • Dorsal stripe (if present) is various colors, with regular edges, and extends to tip of tail (but it may break into patches).
• May have tiny dots of stripe color on sides and large colored area on tops of upper legs.
.... **16**

16

16a • Body has about 16 costal grooves in total ($2^1/_2$ to $5^1/_2$ intercostal folds between toes of adpressed legs).
• Usually less than 55 mm SVL.
• Dorsal stripe is usually red or yellow; it extends to tip of tail; it is irregularly scattered with dark patches on body; it is broad, with solid color on tail.
• Large area of stripe color usually occurs on tops of upper legs.

.... **Western Redback Salamander**

16b • Body has about 17 or more costal grooves in total (four or more intercostal folds between toes of adpressed legs).
• Often more than 60 mm SVL.
• Dorsal stripe is brown to red; it does not extend to tip of tail; it has **either** no dark marks **or** a central zone of dark flecks; it is narrow and fades out or breaks into patches before tip of tail.
• Legs usually do not have any stripe color.
.... **17**

17

17a • Body has about 17 costal grooves (or 4 to $5^1/_2$ intercostal folds between toes of adpressed legs).
• Color is pinkish tan or gray, with conspicuous sprinkling of white flecks or small patches on head, sides and back, but few on underside.
.... **Siskiyou Mountains Salamander**

17b • Body is very long, with 17 to 20 costal grooves (or $6^1/_2$ to $7^1/_2$ intercostal folds between toes of adpressed legs).
• Color is brownish gray, black or reddish, with few or no white flecks on upper surfaces, but more on underside (especially on juveniles).
.... **Del Norte Salamander**

KEY TO HATCHLING TADPOLES

This stage only lasts for a few days, and in some species only a few hours.
If the gills are not evident on your tadpole, also check the Key to Tadpoles.
Some characteristics of very small tadpoles with covered gills are included in the
Key to Tadpoles (p. 138), particularly for the Pacific Treefrog, Bullfrog and Green Frog.
See the glossary of color and markings (p. 171).

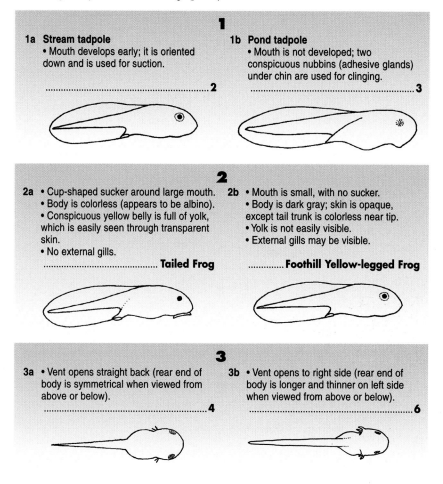

1

1a Stream tadpole
- Mouth develops early; it is oriented
down and is used for suction.

.... **2**

1b Pond tadpole
- Mouth is not developed; two
conspicuous nubbins (adhesive glands)
under chin are used for clinging.

.... **3**

2

2a • Cup-shaped sucker around large mouth.
• Body is colorless (appears to be albino).
• Conspicuous yellow belly is full of yolk,
which is easily seen through transparent
skin.
• No external gills.

.... **Tailed Frog**

2b • Mouth is small, with no sucker.
• Body is dark gray; skin is opaque,
except tail trunk is colorless near tip.
• Yolk is not easily visible.
• External gills may be visible.

.... **Foothill Yellow-legged Frog**

3

3a • Vent opens straight back (rear end of
body is symmetrical when viewed from
above or below).

.... **4**

3b • Vent opens to right side (rear end of
body is longer and thinner on left side
when viewed from above or below).

.... **6**

4

4a • Head is triangular, appears to be separated from belly, with a prominent, narrow ridge connecting to tail (top view).
• Color is pale tan or gray-tan.
.... **Great Basin Spadefoot**

4b • Head does not appear to be separated from belly, and ridge is not noticeable (top view).
• Color is black, charcoal or dark maroon.
.... **5**

5

5a • Dorsal fin is translucent, dark and grainy.
• Tail trunk and body are uniformly dark.
.... **Western Toad**

5b • Dorsal fin is colorless.
• Belly and underside of tail trunk are lighter in color than back and sides.
.... **Woodhouse's Toad**

6

6a • Eyes are conspicuously dark and modify rounded body outline.
.... **7**

6b • Eyes are not developed, do not interrupt rounded body outline.
.... **8**

7

7a • Color is tan; adhesive glands are inconspicuous.
• Round belly is at least as long and wide as head (top view).
.... **Pacific Treefrog**

7b • Color is dark gray; adhesive glands are black and conspicuous.
• Round belly is shorter and narrower than head (top view).
.... **Striped Chorus Frog**

8

8a • Gills may not be visible externally.
• Often less than 4 mm SVL.
.... **Bullfrog** and **Green Frog**

8b • Gills are visible externally after hatching.
• Usually more than 4 mm SVL.
.... **9**

9

9a • Tail length is short, 1½ times body length or less (top view).
.... **10**

9b • Tail length is long, more than 1½ times body length (top view).
.... **12**

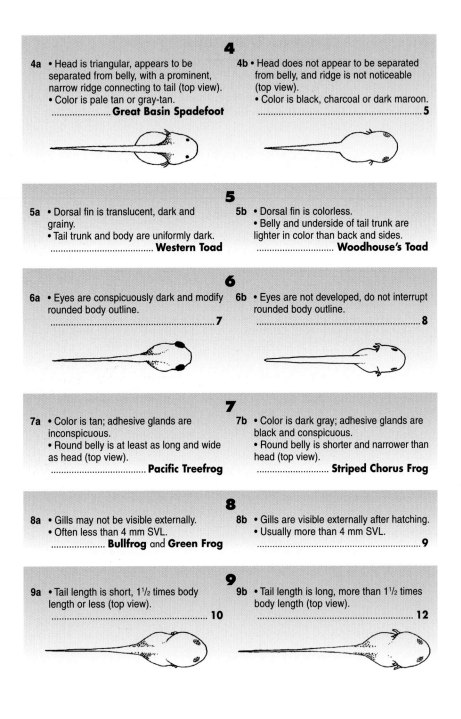

10

10a • Dorsal fin is tall and begins up on middle of back (side view).
  • Gills are mere nubbins, barely visible.
  .... **Red-legged Frog***

10b • Dorsal fin is low and begins near base of tail (side view).
  • Gills are easily visible.
  .... **11**

11

11a • Head is large, broader than body (top view).
  • Gills are distinctly shorter than head length (top view).
  .... **Northern Leopard Frog**

11b • Head is small, narrower than body (top view).
  • Gills appear to be as long as head length (top view).
  .... **Wood Frog**

12

12a • Dorsal fin is low, giving long, streamlined look (side view).
  • Top edge of dorsal fin angles slightly up from back.
  • Dorsal fin is almost opaque, dark charcoal.
  .... **Cascades Frog***

12b • Dorsal fin is tall, making long tail look huge (side view).
  • Top edge of dorsal fin arches steeply up from back.
  • Dorsal fin is translucent, light gray.
  .... **Spotted Frog***

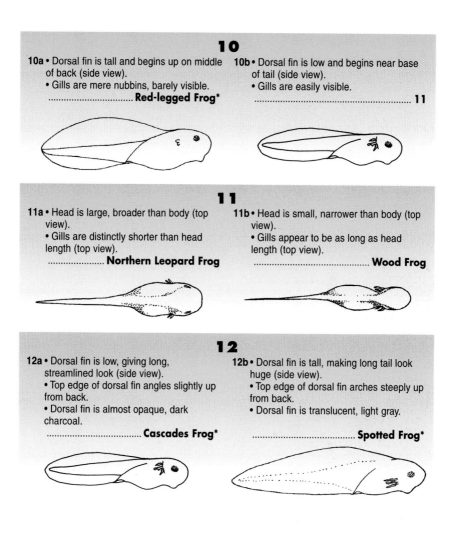

* Also see Confusing Species Comparisons (p. 149).

KEY TO TADPOLES

See the glossary of color and markings (p. 171).

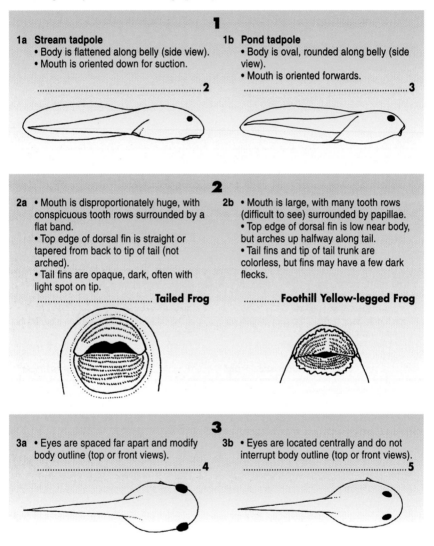

1

1a Stream tadpole
  • Body is flattened along belly (side view).
  • Mouth is oriented down for suction.

.... **2**

1b Pond tadpole
  • Body is oval, rounded along belly (side view).
  • Mouth is oriented forwards.

.... **3**

2

2a • Mouth is disproportionately huge, with conspicuous tooth rows surrounded by a flat band.
  • Top edge of dorsal fin is straight or tapered from back to tip of tail (not arched).
  • Tail fins are opaque, dark, often with light spot on tip.
.... **Tailed Frog**

2b • Mouth is large, with many tooth rows (difficult to see) surrounded by papillae.
  • Top edge of dorsal fin is low near body, but arches up halfway along tail.
  • Tail fins and tip of tail trunk are colorless, but fins may have a few dark flecks.
.... **Foothill Yellow-legged Frog**

3

3a • Eyes are spaced far apart and modify body outline (top or front views).
.... **4**

3b • Eyes are located centrally and do not interrupt body outline (top or front views).
.... **5**

4

4a • Belly is longer and wider than head (top view).
• Nostrils are inconspicuous.
• May reach 24 mm SVL.
.... **Pacific Treefrog**

4b • Belly is shorter and narrower than head (top view).
• Nostrils are large and conspicuous.
• Less than 14 mm SVL.
.... **Striped Chorus Frog**

5

5a • Vent opens straight back.
• Lateral lines are indistinct.
....**6**

5b • Vent opens to right side.
• Lateral lines are distinct on large tadpoles.
....**8**

6

6a • Spiracle is low, touches belly outline (side view).
• Gold flecks on back, arranged in Y-shaped mark.
• [Very small tadpole has triangular head that appears to be separate from body (top view).]
.... **Great Basin Spadefoot**

6b • Spiracle is high on middle of side (side view).
• May have gold flecks on back, but not in Y-shaped mark.
....**7**

7

7a • Color is black or charcoal, may grade to gray on belly.
• Tail trunk is uniformly dark.
• Dorsal fin is translucent, charcoal or densely covered with black flecks.
.... **Western Toad**

7b • Color is dark, with light (white to pale gold) flecks on back and many large, light blotches on belly.
• Tail trunk is dark above, white below.
• Dorsal fin is colorless or very pale gray.
.... **Woodhouse's Toad**

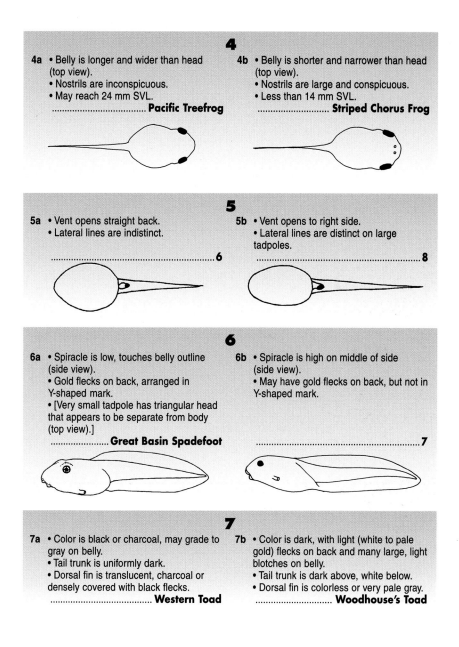

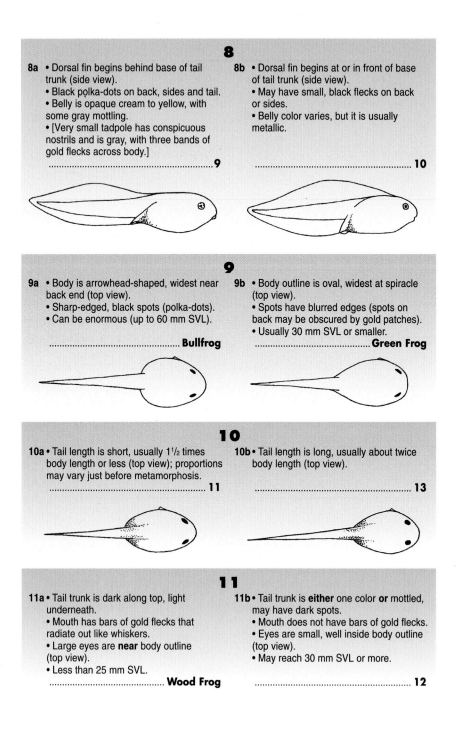

8

8a • Dorsal fin begins behind base of tail trunk (side view).
• Black polka-dots on back, sides and tail.
• Belly is opaque cream to yellow, with some gray mottling.
• [Very small tadpole has conspicuous nostrils and is gray, with three bands of gold flecks across body.]
....**9**

8b • Dorsal fin begins at or in front of base of tail trunk (side view).
• May have small, black flecks on back or sides.
• Belly color varies, but it is usually metallic.
....**10**

9

9a • Body is arrowhead-shaped, widest near back end (top view).
• Sharp-edged, black spots (polka-dots).
• Can be enormous (up to 60 mm SVL).
.............................**Bullfrog**

9b • Body outline is oval, widest at spiracle (top view).
• Spots have blurred edges (spots on back may be obscured by gold patches).
• Usually 30 mm SVL or smaller.
.............................**Green Frog**

10

10a • Tail length is short, usually 1½ times body length or less (top view); proportions may vary just before metamorphosis.
....**11**

10b • Tail length is long, usually about twice body length (top view).
....**13**

11

11a • Tail trunk is dark along top, light underneath.
• Mouth has bars of gold flecks that radiate out like whiskers.
• Large eyes are **near** body outline (top view).
• Less than 25 mm SVL.
.............................**Wood Frog**

11b • Tail trunk is **either** one color **or** mottled, may have dark spots.
• Mouth does not have bars of gold flecks.
• Eyes are small, well inside body outline (top view).
• May reach 30 mm SVL or more.
....**12**

12

12a • Height of dorsal fin is taller than thickness of tail trunk at its base (side view).
• Top edge of dorsal fin arches steeply up from middle of back, above or in front of spiracle (side view).
• Dorsal fin often has gold tone with light dots.
• Belly is opaque (guts are not visible) and has bright gold blotches.

.... **Red-legged Frog***

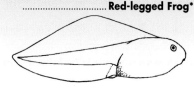

12b • Height of dorsal fin is equal to or less than thickness of tail trunk at its base (side view).
• Top edge of dorsal fin angles slightly up from near base of tail, behind spiracle (side view).
• Dorsal fin may have a few fine dark or light flecks.
• Belly **either** has silver blotches **or** is transparent (so that guts are easily visible).

.... **Northern Leopard Frog**

13

13a • Height of dorsal fin is equal to or less than thickness of tail trunk at its base (side view).
• Top edge of dorsal fin angles slightly up from lower back, behind spiracle (side view).
• Dorsal fin is dark, translucent, densely speckled with fine black or dark gray flecks.
• Body is dark charcoal to black, with fine silver speckling; belly is white or silver.

.... **Cascades Frog***

13b • Height of dorsal fin is taller than thickness of tail trunk at its base (side view).
• Top edge of dorsal fin arches steeply up from near base of tail (side view).
• Dorsal fin is colorless or pale, with a few scattered dark and light flecks.
• Body is brown or gray, with pale gold speckling; belly is pale gold.

.... **Spotted Frog***

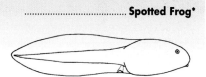

* Also see Confusing Species Comparisons (p. 150).

KEY TO FROGS AND TOADS

See the glossary of color and markings (p. 171).

1

1a 'Toads'
- Skin is dry and lumpy or warty.
- One or two conspicuous hard knobs on heel of hind foot (used for digging).
....2

1b Frogs
- Skin is moist and **either** shiny and smooth **or** grainy.
- No hard knobs on heel of hind foot.
....4

2

2a • One black, spade-shaped knob on heel of hind foot.
- No parotoid glands.
- Pupil is vertical.
- Snout is upturned, with raised nostrils (side view).
.... **Great Basin Spadefoot**

2b • Two knobs on heel of hind foot.
- Parotoid glands are conspicuous.
- Pupil is horizontal.
- Snout is blunt (side view).
....3

3

3a • Parotoid glands are long and narrow.
- Two L-shaped ridges on top of head between eyes.
- Two knobs, one spade-shaped and one round, on heel of hind foot.
.... **Woodhouse's Toad**

3b • Parotoid glands are oval.
- No L-shaped ridges on top of head.
- Two round knobs on heel of hind foot.
.... **Western Toad**

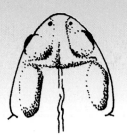
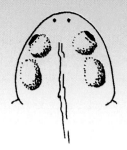

4

4a • Toes are long and straight, with round pads at tips (may be very small).
• Black mask is a sharp-edged, narrow stripe from end of snout at least to shoulder.
....5

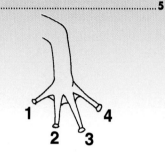

4b • Toes are tapered, rounded or pointed, with no pads at tips.
• May have a dark mask, usually only from eye to ear drum.
....6

5

5a • Toe pad is wider than toe.
• Hind leg is long (lower leg length is about ½ of SVL).
• Mask is a black stripe from end of snout to shoulder.
.... **Pacific Treefrog**

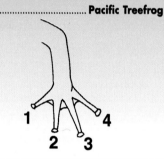

5b • Toe pad is same width as toe or narrower.
• Hind leg is short (lower leg length is about ⅓ of SVL).
• Mask is a black stripe from end of snout along side, often to groin.
.... **Striped Chorus Frog**

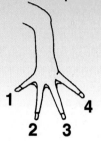

6

6a • Outer toes of hind foot are flatter and wider than other toes.
• Pupil is vertical.
.... **Tailed Frog**

6b • Outer toes of hind foot are no wider than other toes.
• Pupil is horizontal.
....7

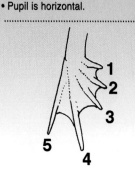

7

7a • Ear drum is conspicuous and is at least as large as eye.
• Prominent groove or fold curves around top of ear drum.
• Color is green to dark brown, with black polka-dots.
• Upper lip is bright green.
.... **8**

7b • Ear drum is inconspicuous and is smaller than eye.
• No groove or fold around ear drum.
• Color is varied, with other distinctive patterns.
• Upper lip may have cream or gold lip line.
.... **9**

8

8a • Dorsolateral folds are distinct but short.
• Bright green lip line extends from snout to shoulder.
• Usually less than 100 mm SVL.
• Eardrum has yellow or gold 'bull's-eye.'
.... **Green Frog**

8b • No dorsolateral folds.
• Bright green on lip is not in distinct line and usually does not extend past ear.
• May be huge, up to 200 mm SVL.
• Eardrum is usually same color as body, or has dark speckling around edge.
.... **Bullfrog**

9

9a • Skin is coarsely grainy (looks like rocks).
• Dorsolateral folds are not apparent.
.... **Foothill Yellow-legged Frog**

9b • Skin is smooth or has scattered small bumps.
• Dorsolateral folds are visible, but may be inconspicuous.
.... **10**

10

10a • Sides of head from eyes to tip of snout have rounded outline (top view).
• Underside of thigh is translucent red to yellow.
• Back is uniform in color, may have black markings.
.... **11**

10b • Sides of head from eyes to tip of snout have long, straight-sided outline (top view).
• Underside of thigh is opaque white.
• Back has pattern of contrasting colors running full length of body.
.... **13**

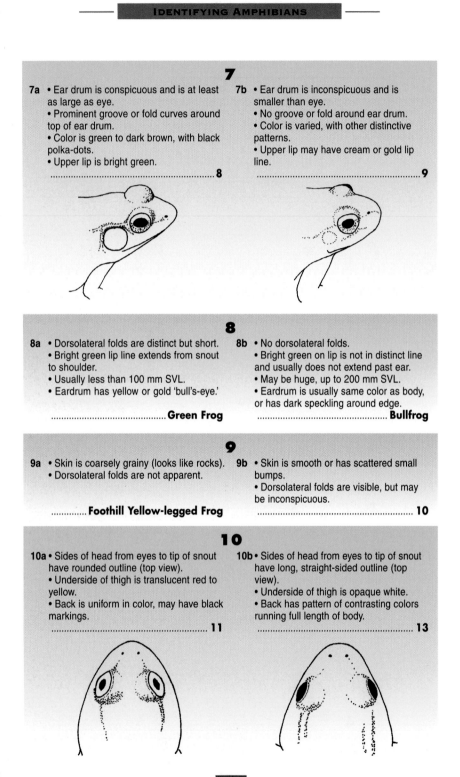

11

11a • Eyes are oriented upwards (top view).
• Hind leg is short (lower leg is shorter than ¹/₂ of SVL).
• Webbing on hind feet is full, almost to tips of toes along both sides of each toe.
• Back has large black spots with blurred or scalloped edges.

.... **Spotted Frog***

11b • Eyes are oriented outwards (top view).
• Hind leg is long (lower leg is longer than ¹/₂ of SVL).
• Webbing is stepped down from inside of one toe to near tip of next (between second and fourth toes).
• Back has **either** small, crisp-edged, black spots or speckling **or** no marks.

.... **12**

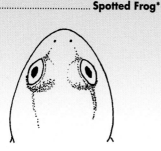
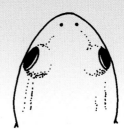

12

12a • Color patches are **larger** in groin than further forward on side.
• Underside of thigh is red.
• Back may have irregular black markings.

.... **Red-legged Frog***

12b • Color patches in groin are **same size** as further forward on side (groin may have greenish wash).
• Underside of thigh is honey-colored, yellow or tan.
• Back may have black spots with crisp edges.

.... **Cascades Frog***

13

13a • Hind leg is long (lower leg length is longer than ¹/₂ of SVL).
• Dorsolateral folds are conspicuously light-colored.
• Back has unmistakable dark oval patches with light borders.

.... **Northern Leopard Frog**

13b • Hind leg is short (lower leg length is equal to or less than ¹/₂ of SVL).
• Dorsolateral folds are same color as back.
• Back **either** is darker than sides **or** has light, central stripe.

.... **Wood Frog**

* Also see Confusing Species Comparisons (pp. 151 & 152).

CONFUSING SPECIES COMPARISONS

NORTHWESTERN SALAMANDER, LONG-TOED SALAMANDER AND ROUGHSKIN NEWT: HATCHLINGS

NORTHWESTERN SALAMANDER	LONG-TOED SALAMANDER	ROUGHSKIN NEWT
Hatches in early to mid spring; develops slowly; may be found in mid spring to mid summer.	Hatches in winter to early spring; develops quickly; only found in early to mid spring.	Hatches in mid spring through summer; develops quickly, but may be found in late spring to late summer.
Head is small, only slightly wider than body; head length is about one-fifth of TL (top view).	Head is large, distinctly wider than body; head length is about one-third of TL (top view).	Head is large, distinctly wider than body; head length is about one-fifth of TL (top view).
Snout is broad; eyes look out to side (top view).	Snout is broad; eyes look out to side (top view).	Snout is narrow; eyes look forward, because front corners are closer together than back corners (top view).
Mouth is straight (front view).	Mouth is straight (front view).	Mouth turns down at corners in a pout (front view).
Gills are usually held out at 45° from body (top view).	Gills are usually held out at 90° from body (top view).	Gills are usually held out at 90° from body (top view).
Gill stalk has few side filaments that occur along full length.	Gill stalk has few side filaments that occur only near base, leaving a long spike.	Gill stalk has few side filaments that occur along full length.
Balancers are short and pale.	Balancers are long, often black-tipped.	Balancers are short and pale.
Upper part of front leg is shorter than lower part; feet are usually hidden beneath gills (top view).	Upper part of front leg is shorter than lower part; feet are usually hidden beneath gills (top view).	Upper part of front leg is longer than lower part; feet are usually visible beyond gills (top view).
No yellow dots along sides.	No yellow dots along sides.	Often has one or two rows of yellow dots along each side.
Hind legs generally appear at 25 mm TL or more.	Hind legs generally appear at 25 mm TL or less.	Hind legs generally appear at 15 mm TL or less.

NORTHWESTERN SALAMANDER, LONG-TOED SALAMANDER AND ROUGHSKIN NEWT: LARVAE

NORTHWESTERN SALAMANDER	LONG-TOED SALAMANDER	ROUGHSKIN NEWT
Develops slowly; may be found at any time of year; generally metamorphoses in second summer, at 40 mm SVL or more; may be neotenic.	Develops quickly; generally found in spring to early summer, except at high elevations; generally metamorphoses by middle of first summer, at 25–35 mm SVL.	Develops quickly; may be found at any time of year; metamorphoses in first or second summer, often at 12–25 mm SVL.
Snout is broad (top view).	Snout is broad (top view).	Snout is narrow (top view).
Mouth is straight (front view).	Mouth is straight (front view).	Mouth turns down at corners in a pout (front view).
Gills are one color.	Gills are one color.	Gills have dark speckles, often droop down at tip.
Side filaments are long and willowy (gills appear full, like ostrich plumes), occur along entire gill stalk.	Side filaments are graduated in length (longer near base of gill), occur near base of gill stalk, leaving a long spike at tip.	Side filaments are of uneven lengths (gills appear ragged), occur along entire gill stalk.
Poison glands occur on parotoid area and along tail ridge, become visible at more than 40 mm SVL.	Poison glands are not apparent.	Poison glands are not apparent.
May have one row of yellow dots low along sides.	No yellow dots along sides.	Two or three rows of yellow dots high along sides.
Underside is white or gray.	Underside is white or gray.	Underside usually has an orange or pink tinge.
Eyes are dark.	Eyes are gold.	Eyes are gold, often with a dark bar across them.
Body has chunky build.	Body often has slender build.	Body has slender build.
Legs and feet are wide.	Legs and feet are slender.	Legs and feet are slender.
Skin is usually opaque.	Skin is usually translucent.	Skin is usually opaque.
Olive or brown in color, often with large dark spots on back and fin.	Tan in color, with a fine network of tiny, dark speckling.	Gold or brown in color, with dense, dark speckling.

COPE'S GIANT SALAMANDER AND PACIFIC GIANT SALAMANDER: LARVAE

These two species are most difficult to distinguish when they are less than 40 mm SVL. When Pacific Giant Salamanders reach about 120 mm SVL, they may not retain the proportions that differentiate them from Cope's Giant Salamanders.

COPE'S GIANT SALAMANDER	PACIFIC GIANT SALAMANDER
Long build, but generally no more than 100 mm SVL.	Chunky build, may be huge (up to 190 mm SVL).
Toes of adpressed legs do not touch.	Toes of adpressed legs touch or overlap.
Gill filaments are shorter than gill stalks, so gills appear separate (like larch needle bundles or shaving brushes).	Gill filaments are longer than gill stalks and frequently cover them, so gills appear as one bush on each side of head.
Head is long, flat, rectangular (about as narrow at gills as at eyes) and almost as narrow as body; head length is about one-quarter of SVL (top view).	Head is massive, wedge-shaped (wider at gills than at eyes, but more rounded on huge larvae) and much wider than body; head length is about one-third of SVL (top view).
Dorsal fin is low in height; begins behind vent; top edge slopes up from midway down tail (side view); often has light or bright gold markings, especially along top edge.	Dorsal fin is tall; begins in front of vent (often in front of hind legs); top edge arches up from near base of tail (side view); usually has black markings, especially along top edge and at tip.
Dorsal color is brown or gray-brown, often with tan or gold glandular patches that may appear finely pitted.	Dorsal color is brown or gray-brown; may have light mottling on back or sides (but usually does not appear pitted).
Underside is usually lavender gray; often has a whitish throat, vent or narrow central stripe.	Underside is usually whitish; may have gray mottling with a white vent.

RED-LEGGED, CASCADES AND SPOTTED FROGS: HATCHLINGS

RED-LEGGED FROG	CASCADES FROG	SPOTTED FROG
Overall stubby appearance from a short tail and a tall dorsal fin.	Overall streamlined appearance from a long tail and a low dorsal fin; dark.	Overall sweeping appearance from a long tail and a tall dorsal fin.
Tail length is short, no more than 1½ times body length (top view).	Tail length is long, more than 1½ times body length (top view).	Tail length is long, more than 1½ times body length (top view).
Dorsal fin is translucent, light gray.	Dorsal fin is nearly opaque, charcoal.	Dorsal fin is translucent, light gray.
Top edge of dorsal fin arches steeply up from middle of back (side view).	Top edge of dorsal fin angles slightly up from near base of tail (side view).	Top edge of dorsal fin arches steeply up from near base of tail (side view).
Tail trunk angles up slightly from body (side view).	Tail trunk does not angle up from body (side view).	Tail trunk does not angle up from body (side view).
Gills are mere nubbins, barely visible.	Gills are long, like gnarled fingers.	Gills are long, like gnarled fingers.
Hatchling clings to egg mass, or to nearby vegetation if disturbed.	Hatchling often found in water pooled on top of stranded egg mass.	Hatchling often found in water pooled on top of stranded egg mass.

RED-LEGGED, CASCADES AND SPOTTED FROGS: TADPOLES

These features may vary just before metamorphosis, at which time the juvenile features may be more useful in distinguishing between these species.

RED-LEGGED FROG	CASCADES FROG	SPOTTED FROG
Overall stubby appearance from short tail and tall dorsal fin.	Overall streamlined appearance from long tail and short dorsal fin; dark.	Overall sweeping appearance from long tail and tall dorsal fin.
Tail length is short, usually no more than 1½ times body length (top view).	Tail length is long, usually about twice body length (top view).	Tail length is long, usually about twice body length (top view).
Top edge of dorsal fin arches steeply up from middle of back, above or in front of spiracle (side view).	Top edge of dorsal fin angles slightly up from lower back near base of tail, behind spiracle (side view).	Top edge of dorsal fin arches steeply up from base of tail, behind spiracle (side view).
Dorsal fin is taller than thickness of tail trunk (side view).	Dorsal fin is equal to or less than thickness of tail trunk (side view).	Dorsal fin is taller than thickness of tail trunk (side view).
Body is long, oval (top view).	Body is egg-shaped (top view).	Body is egg-shaped (top view).
Dorsal fin has a golden tone, with light and gold dots, or it is colorless.	Dorsal fin is dark, translucent, densely speckled with black or charcoal.	Dorsal fin is pale or colorless, with a few scattered dark and light flecks.
Body color is tan, with bright gold or brassy blotches, especially on underside.	Body color is charcoal to black, with fine, silver (or pale gold) flecks, especially on sides and belly, which looks white or silver.	Body color is brown or gray, with dull gold flecks or speckling, especially on sides; belly is pale gold.
Small tadpole has gold flecks in two vague lines along back.	Small tadpole is very dark, with scattered silver flecks mostly on belly.	Small tadpole is dark, with scattered pale gold flecks mostly on belly.

RED-LEGGED, CASCADES AND SPOTTED FROGS: JUVENILES

RED-LEGGED FROG	CASCADES FROG	SPOTTED FROG
Eyes are gold, oriented to side (top view).	Eyes are gold, oriented to side (top view).	Eyes are gold, yellow or dark, oriented upwards (top view).
Dorsolateral folds are distinct, less so on lower back, and lower back looks wide and round between them.	Dorsolateral folds are distinct on full length of back to hip, and lower back looks angular or flat between them.	Dorsolateral folds are inconspicuous, especially on lower back, which looks very broad and plump.
Snout is short and rounded.	Snout is short and slightly tapered.	Snout is long and tapered.
Eyes do not interrupt body outline (top view).	Eyes do not interrupt body outline (top view).	Eyes may protrude slightly to side (top view).
Lip line is light and short; it is vague on snout.	Lip line is light and distinct to end of snout, making snout appear pointed.	Lip line is light; it may be vague or distinct on snout.
Underside of thigh is pale pink.	Underside of thigh is dull yellow or tan.	Underside of thigh is dull pink.
May have inconspicuous harness-shaped pattern on chest.	May have inconspicuous harness-shaped pattern on chest.	Sharply defined, harness-shaped pattern on chest.

 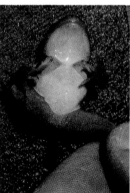

RED-LEGGED, CASCADES AND SPOTTED FROGS: ADULTS

RED-LEGGED FROG	CASCADES FROG	SPOTTED FROG
Eyes are gold, oriented to side (top view).	Eyes are gold, oriented to side (top view).	Eyes are bright yellow, chartreuse or gold, oriented upwards (top view).
Hind legs are long; lower leg length is more than half of SVL.	Hind legs are long; lower leg length is more than half of SVL.	Hind legs are short; lower leg length is less than half of SVL.
Webbing on hind foot is not full—it is stepped down along inside of one toe to near tip of next (between second and fourth toes).	Webbing on hind foot is not full—it is stepped down along inside of one toe to near tip of next (between second and fourth toes).	Webbing on hind foot is full—it extends almost to tips of toes along both sides of each toe.
Dorsolateral folds are distinct to hip.	Dorsolateral folds are distinct to hip.	Dorsolateral folds are inconspicuous, especially on lower back.
Groin color patches are larger than those further forward on sides.	Groin color pattern is pale, or similar to that further forward on sides, and may have a green wash.	Groin color and pattern are similar to that further forward on sides, or they are plain gray.
Back has black speckling or irregular marks (or no marks).	Back has round or angular black spots with crisp edges, sometimes with a light center; may have no spots.	Back has huge black spots with blurred or scalloped edges, generally with a light bump in center.
Underside of entire hind leg is usually translucent red.	Underside of thigh is translucent yellow or tan.	Underside of thigh is opaque, with a red or orange surface color.
Snout is rounded; eyes do not interrupt outline (top view).	Snout is rounded; eyes do not interrupt outline (top view).	Snout is slightly tapered; eyes may protrude slightly to side (top view).
Mask behind eye is vague, black or brown, generally sprinkled with light dots and patches.	Mask behind eye is distinct, brown or black.	Mask behind eye is often absent, or it is pale or has scattered light patches, especially on ear drum.
Lip line is light; it is vague or absent on snout.	Lip line is light; it is distinct to below nostril or to end of snout.	Lip line is light; it is either distinct or blurred on snout.
Dark brown (often with a red tone) to light pink-tan or golden tan in color.	Dark brown (may have green tone) to golden tan in color.	Dark brown (with red or green tone) to golden tan in color.
Throat, chest and sides of belly often have a fine speckling of gray or red dots.	Chest and belly are usually white or yellow, with no markings; throat is white, often with a large, pale gray mottling.	Throat and chest are white or have gray mottling; sides of belly have red, orange or sometimes gray mottling; center of belly has gray mottling or white; may show triangles from arms onto chest.

FINDING AMPHIBIANS

WHERE TO LOOK

To find amphibians in the Pacific Northwest you will have to consider both aquatic and terrestrial habitats, because an amphibian's habitat preferences change with the seasons and in different development stages. Also, be sure to look for amphibians when they are most likely to be active at the surface in your area (see Amphibian Activity through the Seasons, p. 156).

TYPICAL HABITATS

We have identified nine general amphibian habitat types for this region (see Amphibian Habitats, p. 26). These habitats provide seasonally usable conditions for several development stages of amphibians, including habitats for foraging, reproduction and just hanging out. The table on page 155 lists the species likely to be found in these habitats (during appropriate seasons) in Oregon, Washington and British Columbia. Keep in mind that some species may be found in habitats other than those for which they are listed, particularly as they move to and from their breeding sites.

SPECIAL HABITATS

It has been our experience that certain amphibian species are often found in association with specific plants. The descriptions that follow are of habitats and plant groups that can be useful as indicators that certain amphibians may be present. Other amphibians may also be present, but we focus on these relationships. This section is not meant to restrict your search image but to increase your awareness when you recognize these habitats.

Mountain Wet Meadows are meadows with cold springs, bogs, ponds, elk or bear wallows or very slow streams at mid to high elevations in the Cascade Range. Typical plants include mosses, low-growing grasses or sedges, marsh marigold (*Caltha bicolor*), sundews (*Drosera* spp.) and huckleberries (*Vaccinium* spp.). These meadows are surrounded by conifer forests that are often dominated by western red cedar (*Thuja plicata*) or Engelmann spruce (*Picea engelmannii*). **Key Species:** Northwestern Salamander, Western Toad, Cascades Frog.

Cascades Frog

Forested Wetlands are shaded ponds (often beaver ponds) at low to mid elevations, primarily west of the Cascade Range. They are characterized by slough sedge (*Carex obnupta*) and Oregon ash (*Fraxinus latifolia*), often with a middle layer of willow (*Salix* spp.). These wetlands are often adjacent to moist conifer or hardwood forests, although we have found them in urban areas where they are secluded from disturbance. **Key Species:** Northwestern Salamander, Red-legged Frog.

Red-legged Frog eggs

Warmwater Ponds and Emergent Wetlands are open, sparsely shaded water at low elevations. Cattail (*Typha latifolia*), or bulrush (*Scirpus* spp.) and duckweed (*Lemna minor*) characteristically occur in these wetlands, which also include non-native plants that are adapted to disturbed conditions. These systems once provided habitat for the Spotted Frog. **Key Species:** Bullfrog.

Bullfrog

Headwater Streams are streams that occur at most elevations in the Cascade Range and Coast Mountains. They are clear, cold and fast-flowing, and their substrate is of cobbles or small boulders. They typically occur in dense conifer forests with large downed logs, devil's club (*Oplopanax horridus*), liverworts and various members of the saxifrage family. In our experience, if you find one of these species, the others are probably present as well. **Key Species:** Cope's Giant Salamander, Tailed Frog, Torrent Salamanders.

Cope's Giant Salamander

SPECIES OCCURRENCE BY TYPICAL HABITATS

HABITAT	West of or in Cascade Range or Coast Mountains	On both sides of Cascade Range or Coast Mountains	East of Cascade Range or Coast Mountains
MARSH AND WET MEADOW	Northwestern Salamander Roughskin Newt Red-legged Frog Cascades Frog	Long-toed Salamander Western Toad Pacific Treefrog Spotted Frog[†] Bullfrog	Woodhouse's Toad[†] Striped Chorus Frog[†] Wood Frog Northern Leopard Frog[†]
SHALLOW POND	Northwestern Salamander Roughskin Newt Red-legged Frog Cascades Frog	Long-toed Salamander Western Toad Pacific Treefrog Spotted Frog[†] Bullfrog	Tiger Salamander[†] Woodhouse's Toad[†] Great Basin Spadefoot Striped Chorus Frog[†] Wood Frog Northern Leopard Frog[†]
DEEP POND	Northwestern Salamander Roughskin Newt Red-legged Frog Cascades Frog	Spotted Frog[†] Bullfrog Green Frog[†]	Tiger Salamander[†] Northern Leopard Frog[†]
SMALL STREAM	Pacific Giant Salamander Torrent Salamanders Dunn's Salamander[*] Van Dyke's Salamander[†*] Red-legged Frog[*] Cascades Frog[*] Foothill Yellow-legged Frog	Cope's Giant Salamander[†] Western Toad[*] Tailed Frog	Pacific Treefrog[*]
MEDIUM STREAM	Pacific Giant Salamander Torrent Salamanders[*] Dunn's Salamander[*] Van Dyke's Salamander Red-legged Frog[*] Cascades Frog[*] Foothill Yellow-legged Frog	Cope's Giant Salamander[†] Western Toad[*] Tailed Frog Spotted Frog[†]	Pacific Treefrog[*]
LARGE STREAM	Pacific Giant Salamander Dunn's Salamander[*] Foothill Yellow-legged Frog	Western Toad[*] Tailed Frog Bullfrog	Pacific Treefrog[*]
FOREST AND LOGS	Northwestern Salamander Roughskin Newt Pacific Giant Salamander Clouded Salamander Black Salamander[†] Oregon Slender Salamander California Slender Salamander Ensatina Larch Mountain Salamander[†] Van Dyke's Salamander[†] Western Redback Salamander Del Norte Salamander[†] Siskiyou Mountains Salamander	Western Toad Pacific Treefrog	
TALUS	Pacific Giant Salamander Black Salamander[†] Oregon Slender Salamander Ensatina Dunn's Salamander Larch Mountain Salamander[†] Van Dyke's Salamander[†] Western Redback Salamander Del Norte Salamander[†] Siskiyou Mountains Salamander[†]		Coeur d'Alene Salamander[†]
GRASS AND SHRUB		Long-toed Salamander Western Toad Pacific Treefrog	Tiger Salamander[†] Woodhouse's Toad[†] Great Basin Spadefoot Northern Leopard Frog[†]

[†] May occur at only a few localities.
[*] Occurs on bank or in stream edge.

AMPHIBIAN ACTIVITY THROUGH THE SEASONS

Amphibians, which respond strongly to changes in temperature and moisture, are most likely to be active at the surface during rainy warming trends following cold weather and during rainy cooling trends following dry or hot weather. There are many stories of vast nighttime parades of salamanders or frogs during the first warm rains of fall or spring, and these movements may introduce the breeding season. In most pond-breeding species, the males arrive at the breeding site earlier and remain longer than the females, which stay just long enough to lay their eggs. Some fully terrestrial salamanders may mate in the fall, but they do not lay their eggs until spring.

The definitions of seasons and the timing of seasonal activities that follow are general, and they should be tailored to different elevations, latitudes, regional climate variations and unexpected annual weather patterns. For example, early spring might mean mid-February in the Willamette Valley and late April in the Blue Mountains or Peace River country.

WINTER
(Some periods of freezing weather; often December to mid February.)

During periods of warm rain:

- Long-toed Salamanders gather at breeding sites and lay their eggs. Sudden, prolonged, sub-freezing weather can kill both adults and eggs.
- The first Red-legged Frogs lay their eggs.
- The low croaking of Pacific Treefrogs can be heard away from ponds, but they are not yet breeding.

EARLY SPRING
(Frost some nights, some sun, some late snows; often late February and March.)

During periods of warm rain:

- The first Northwestern Salamanders lay their eggs.
- The last Long-toed Salamanders lay their eggs, and many of the earlier eggs hatch.

- The last Red-legged Frogs lay their eggs.
- Pacific Treefrogs begin chorusing and may begin to lay eggs.
- Most Cascades Frogs and the first Spotted Frogs lay their eggs.
- Wood Frogs and Striped Chorus Frogs call during the lengthening daylight hours and lay their eggs.

MID SPRING
(Early flowers are in bloom; frost some nights; often late March and April.)

- Fully terrestrial salamander species are active and many are laying their eggs.
- The last of the Long-toed Salamander eggs hatch.
- Many Northwestern Salamanders and the first Roughskin Newts move to breeding ponds and lay their eggs.

- Many Pacific Giant Salamanders and Cope's Giant Salamanders probably breed during the spring, but they are active in streams all year.
- Tiger Salamanders breed at this time.
- Pacific Treefrogs chorus constantly and lay their eggs, and the earliest eggs hatch.

Toads in amplexus

- Western Toads gather at breeding ponds and lay their eggs, which hatch quickly.
- Most Red-legged and Cascades frogs eggs hatch, and the first Spotted Frog eggs hatch.
- Some Bullfrogs emerge from hibernation during warm days.
- The first Great Basin Spadefoots emerge, and during warm, rainy periods they lay their eggs, which hatch within a few days.
- Most Spotted Frogs and Northern Leopard Frogs breed at this time.
- Wood Frog and Striped Chorus Frog tadpoles are growing rapidly.

LATE SPRING
(Many flowers are in bloom; some warm days and nights; often late April and May.)

- Fully terrestrial salamander species probably breed at this time.
- Many Roughskin Newts lay their eggs, and the earliest eggs hatch.
- Northwestern and Tiger salamander eggs hatch.
- Torrent Salamanders probably breed at this time, but they may breed over an extended period.
- The earliest Long-toed Salamander larvae metamorphose.

- All frogs are active.
- Pacific Treefrogs are still chorusing and breeding, and some eggs are hatching.
- Western Toads at cold-water sites and Woodhouse's Toads lay their eggs, which hatch within a few days.
- Bullfrogs and Green Frogs are active, and they may be breeding in very warm ponds.
- Red-legged, Cascades and Spotted frog tadpoles are growing.
- Most Great Basin Spadefoots emerge, and during rainy periods they lay their eggs, which hatch within a few days.
- Northern Leopard Frog eggs hatch.
- Foothill Yellow-legged Frogs breed as the stream flows lessen.
- Striped Chorus Frog tadpoles begin to metamorphose.

EARLY SUMMER
(Some periods of hot days; sometimes late May, often June and early July.)

- All larval salamanders are active.
- Many Roughskin Newts are still laying eggs, and earlier ones hatch. Some adult newts stay in the ponds for most of the spring and summer.
- Most Long-toed Salamander larvae metamorphose.

Pacific Treefrog eggs

- All tadpoles are actively foraging and growing fast.
- Pacific Treefrogs, Bullfrogs, and Green Frogs are still laying their eggs, which hatch quickly.
- Foothill Yellow-legged Frogs and Tailed Frogs lay eggs.
- Most tadpoles of Great Basin Spadefoots and Woodhouse's Toads, and early Red-legged and Spotted frogs, metamorphose.
- Green Frog tadpoles that are yearlings (one winter as a tadpole) and Bullfrog tadpoles, which may be several winters old, metamorphose.
- Wood Frog tadpoles metamorphose.

LATE SUMMER
(The hottest, driest days; often late July to early September.)

Long-toed Salamanders and Western Toads

- The last Long-toed Salamander larvae and the first Northwestern and Tiger salamander larvae metamorphose and hide in the mud or under moist debris beside ponds.
- The last Foothill Yellow-legged Frog eggs may be hatching, while the earliest tadpoles are metamorphosing.
- Most tadpoles of Western Toads, Pacific Treefrogs, Cascades Frogs and Northern Leopard Frogs metamorphose, as do the last tadpoles of Woodhouse's Toads, Red-legged Frogs and Spotted Frogs.
- Froglets and toadlets of most species are commonly encountered as they disperse.
- The last Bullfrogs are still laying eggs, which hatch in a few days.
- Adult Red-legged Frogs and other frogs and toads may move to permanent streams or moist forest sites.
- Tailed Frog tadpoles that are several years old metamorphose.

EARLY FALL
(Warm days and cool nights [frost possible]; often late September and October.)

- Many Roughskin Newts, Tiger Salamanders and late Northwestern Salamanders metamorphose.
- Cope's and Pacific Giant Salamanders probably breed at this time.
- Late tadpoles of Foothill Yellow-legged Frogs, Western Toads, Pacific Treefrogs, Cascades Frogs metamorphose.
- Early Green Frog tadpoles metamorphose.
- Many Tailed Frogs probably mate at this time.

LATE FALL
(Heavy rain west of Cascades and Coast Mountains; snow in mountains; usually November.)

- All salamanders are active at the surface in response to fall rains.
- As temperatures drop to near freezing, many metamorphosed forms and fully terrestrial salamanders overwinter underground.
- Some larvae of Northwestern Salamanders, Tiger Salamanders and Roughskin Newts, Long-toed Salamander larvae at high elevations and neotenic forms of some species overwinter in mud or water of permanent ponds.

- Some Long-toed Salamanders move to breeding ponds.
- Many fully terrestrial salamanders mate at this time.
- Roughskin Newts are particularly conspicuous.

- Bullfrog and late Green Frog tadpoles overwinter in ponds.
- Some Cascades Frog and Spotted Frog tadpoles may overwinter in ponds.
- Some Red-legged Frogs move to breeding ponds.

SEARCH TECHNIQUES

There are several different techniques you can use to find amphibians: choose those that will fit with your objectives, the season, the weather, the habitat and the species and development stages of the amphibians you will potentially find. Also be sure to use the techniques that meet these requirements with the least damage to the habitat and its inhabitants—it is important to follow the requirements for protecting critical habitats while you work. Repeat surveys during the best seasons for finding amphibians.

Observing: Search for amphibians where they are resting (on the bottom of a pool, in the grass at the edge of a pond, etc.). Be alert to detect the motion of amphibians as they respond to your presence.

Uncovering: Lift rocks and pieces of wood to find amphibians where they hide, on land, around the edges of ponds or in streams. When working in streams, lift the rocks and gently sift through the gravel with your fingers. A large dipnet or screen can be held or set into the rocks immediately downstream to catch amphibians that float down when disturbed, or you can scoop them up with an aquarium net or D-net. Stream work is easiest and safest when the water level is low (late spring through early fall). **Be sure** to replace rocks, wood and bark in their original positions, and to rebuild disturbed log sections.

Dipnetting: To find larvae, use a large, long-handled net to scoop water and mud from the bottom of ponds and from around submerged vegetation. This works best in spring and summer. Minimize the amount of sediment you stir up.

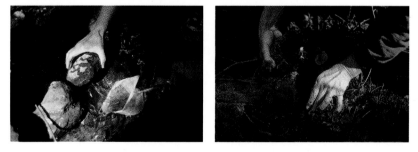

Night-lighting: Search at night with a light. During periods of warm rain, this can frequently be done from a car (on roads with little traffic). It also works well on foot along streams, but this is only recommended in gentle terrain and easily accessible sites.

Listening: Listen for the calls of frogs and toads near potential breeding sites. This technique works best at night. It is most useful on the east side of the Cascade Range and Coast Mountains, because many western species have calls that carry only a short distance. Repeat auditory surveys every few days during the most likely breeding season (see Amphibian Activity through the Seasons, p. 156). Some people use an automated recording system left at the study site. Tapes of frog and toad calls are available for the identification of frogs by sound (Davidson, 1995).

Other search techniques, including pitfall trapping and intensive stream surveys, are not recommended for general use; they should be used only by responsible research scientists. Pitfall traps, which are set in the ground for amphibians to fall into them, must be checked several times each day, because trapped animals can die if water collects in the pit or it dries out. An established form of stream survey that involves removing all the rocks down to bedrock in 10 m sections of stream and using seine nets to capture all the amphibians may require a special permit. These techniques are too expensive and time-consuming for general use.

Make sure nets, boots, plastic bags and all other equipment is washed before moving to a new area, to prevent the transfer of diseases, fungi, weed seeds and other unwanted organisms.

Further information on search techniques is available in several publications listed in the references (p. 172).

PLANNING AMPHIBIAN SURVEYS

Protocols for surveys and monitoring are currently being developed by the Pacific Northwest Amphibian and Reptile Consortium (Leonard, in prep.). Here we briefly describe several general types of surveys, which use one or more of the basic search techniques described above.

A word of caution to wildlife managers of amphibian surveys: Consultants and biologists need to demonstrate that they are knowledgeable about amphibian identification and habitats. Startling new finds need to be confirmed by an expert.

TYPES OF SURVEYS

The type of survey you conduct depends on your specific objectives, the type of site you are surveying and the time and equipment resources available to you.

BASIC SURVEYS AT PLANNED PROJECT SITES

The objectives are to determine what species are present and to obtain some general data on relative abundance.

Before you begin:
- review the published literature about the site
- locate ponds, bogs, talus slopes and other special sites
- determine potential habitats for listed and sensitive species
- understand the seasonal activity of expected species.

Surveys should be completed during the seasons and weather when amphibians are most likely to be active on the surface.

Note: The absence of a species from a particular site cannot be determined, but it can be inferred by repeated and careful surveys during the most appropriate seasons and weather.

> **Helpful hint:** In our experience, a quick way to estimate the water depth at a site on the day that you are surveying it is to measure the height of the water-proof footwear you chose to wear on that day and add 5 cm (2 in).

DATA COLLECTION IN STREAM SURVEYS AND DURING FISH MONITORING

The objective is to collect preliminary data on the presence and relative abundance of amphibian species.

For each stream section surveyed, crews should try to capture, identify and report
a) amphibians found in a regular sampling of measured units and
b) amphibians incidentally observed throughout the section.

Amphibians observed while electro-shocking fish should be netted, identified, recorded, resuscitated and released.

LONG-TERM MONITORING AT SPECIFIC LOCATIONS

The objective is to closely estimate population sizes and reproductive success over many years. Known sites where particular species are consistently observed, especially discrete breeding sites, such as ponds, should be surveyed several times each year during the seasons appropriate for observing different life stages. Use the same search techniques and level of effort consistently.

INCIDENTAL AND CASUAL OBSERVATIONS

Wildlife and fish biologists and naturalists should report breeding activity and uncommon species of amphibians observed during other natural resource surveys (see What Data to Collect, p. 163). Biologists and researchers can do follow up surveys.

INTENSIVE STUDIES OF INDIVIDUAL SPECIES OR ASSOCIATIONS

Intensive studies, particularly those involving mark and recapture or pitfall trapping, are time consuming and destructive to both animals and habitats; only responsible research scientists should undertake them.

AMPHIBIAN SURVEY EQUIPMENT LIST

NECESSARY:

- notebook and pencil
- camera and film
- several heavy duty 1 quart and smaller plastic bags
- wide, plastic tubs
- hand lens (at least 10x), on a string
- one or two aquarium nets, 12–15 cm wide (5–6 in), with a strong handle and duct tape around the rim
- 15 cm (6 in) flexible plastic ruler
- thermometer
- maps
- compass
- general field equipment (whichever 10 essentials list you like)
- general emergency kit
- patience

OPTIONAL:

- test tubes or coin-collectors' vials
- long-handled dipnet, D-net, etc.
- section of coarse screen or net (duct tape around edge)
- close-up lens and flash equipment for camera (necessary for voucher specimens)
- luck

WHAT DATA TO COLLECT

The type and intensity of your survey will determine how much data you should collect. Data forms are available from the Pacific Northwest Declining Amphibian Populations Task Force (see below). Even for incidental observations of amphibians, however, at least the following information should be recorded:

- *Date*
- *Observer*
- *Location:* county, legal description, known landmarks or directions to the site
- *Habitat:* e.g., stream, pond, meadow or other general information
- *Weather:* general conditions, air temperature (at least an estimate)
- *Water temperature* (if known)
- *Species:* list all amphibian species encountered
- *Development stages:* count or estimate the numbers of each species and development stage that you observe.

Data from any full survey, monitoring or casual observation of a species listed by any agency (see Status of Amphibians, p. 165) or found outside its known range should be copied and given to at least one of the following:

- the wildlife biologist in the local office of the land management agency
- the Pacific Northwest Declining Amphibian Populations Task Force

USFS, PNW Research Station   DAPCAN
3200 SW Jefferson Way   #235F, 560 Johnson Street
Corvallis, OR 97331   Victoria, BC V8W 3C6

- the local wildlife diversity biologist, the state office of the Oregon or Washington Department of Fish & Wildlife, or the regional or provincial office of the B.C. Wildlife Branch
- the state natural heritage program

Oregon Natural Heritage Program   Washington Natural Heritage Program
821 SE 14th Ave.   P. O. Box 47016
Portland, OR 97214   Olympia, WA 98504-7016

Juvenile Ensatinas

STREAM ORDERS

The designations 1, 2, 3, etc. refer to stream segments in a drainage basin network, according to the U.S. Forest Service Region 6 classifications. First order streams are the smallest, unbranched, often intermittent tributaries, which end at an outer point or spring. Second order streams are stream segments produced by the junction of two first order streams. Third order streams are stream segments produced by the junction of two second order streams, etc.

There are several other systems for classifying streams, including the U.S. Geological Survey, which uses exactly the opposite numbering.

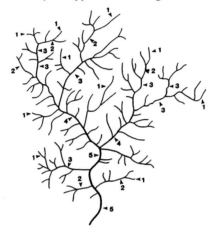

SCIENTIFIC TAKING PERMIT

Because this manual uses only non-damaging identification methods, you may not always successfully differentiate between confusing species. You should carry a camera in the field and take photographs as vouchers, but if you absolutely must identify the species of an individual, and you cannot take photographs to show the distinguishing features, you may need to collect the individual and take it to an expert.

If you are working on amphibians in the Pacific Northwest, state and provincial laws require that you have a permit to collect all species of wildlife that are protected by legal statutes or regulations (see Status of Amphibians, p. 165). A permit for a valid reason (amphibians should never be kept as pets), as well as the latest list of protected species, may be obtained by applying to one of the following agencies:

Oregon Department of Fish and Wildlife
Wildlife Diversity Program
P. O. Box 59
Portland, OR 97207

Washington Department of Fish
and Wildlife
600 Capitol Way, North
Olympia, WA 98501-1091

B.C. Ministry of Environment, Lands and Parks
Wildlife Branch
780 Blanshard Street
Victoria, BC V8V 1X4

STATUS OF AMPHIBIANS

These checklists must be updated periodically.

STATUS OF AMPHIBIANS IN OREGON: JUNE 1996

ODFW: Oregon Department of Fish & Wildlife (as of December 1995)

Pr = Legally protected from taking

S = Sensitive

   c = critical

   p = peripheral, naturally rare

   v = vulnerable

   u = undetermined status.

TNC: The Nature Conservancy (as of December 1995)

G = Global (status throughout its range)

S = Subnational (status within Oregon)

   1 = critically imperiled

   2 = imperiled

   3 = rare or uncommon

   4 = not rare, long-term concern

   5 = widespread, abundant, secure

   u = unknown rank

   E = exotic

   ? = insufficient data.

BLM: Bureau of Land Management, USDI

TS = Tracking species

AS = Assessment species

BS = BLM sensitive

ROD = 'Survey & manage' species described in the Record of Decision for Amendments to USFS and BLM Planning Documents within Range of Northern Spotted Owl, April 1994.

USFS: Forest Service, Region 6, USDA

S = Sensitive

ROD = 'Survey & manage' species described in the Record of Decision for Amendments to USFS and BLM Planning Documents within Range of Northern Spotted Owl, April 1994.

USFWS: Fish & Wildlife Service, USDI (as of February 1996)

C = Candidate, sufficient information (number refers to priority).

SPECIES	ODFW	TNC	BLM	USFS	USFWS
Northwestern Salamander		G5/S5			
Long-toed Salamander		G5/S5			
Tiger Salamander	S/u	G5/Su			
Roughskin Newt (the subspecies near Crater lake)		G5/S1			
Cope's Giant Salamander	Pr-S/u	G3/S1	AS	S	
Pacific Giant Salamander		G5/S5			
Cascade Torrent Salamander	Pr-S/v	G2/S3	TS		
Columbia Torrent Salamander	Pr-S/c	G2/S3			
Southern Torrent Salamander	Pr-S/c	G4/S3			
Clouded Salamander	Pr-S/u	G5/S4			
Black Salamander	Pr-S/p	G4/S2			
Oregon Slender Salamander	Pr-S/u	G3/S3			
California Slender Salamander	Pr-S/p	G5/S2			
Ensatina		G5/S5			
Dunn's Salamander		G4/S5			
Larch Mountain Salamander	Pr-S/v	G2/S2	ROD	ROD	
Western Redback Salamander		G5/S5			
Del Norte Salamander	Pr-S/v	G3/S2	ROD	ROD	
Siskiyou Mountains Salamander	Pr-S/v	G2/S2	ROD	ROD	
Tailed Frog	Pr-S/v	G3/S3	AS		
Great Basin Spadefoot		G5/S5			
Western Toad	S/v	G4/S4			
Woodhouse's Toad	S/p	G5/S2			
Pacific Treefrog		G5/S5			
Red-legged Frog	Pr-S/u	G4/S4	TS	S	
Cascades Frog	Pr-S/v	G4/S3	AS		
Spotted Frog (west coast population)	Pr-S/c	G4/S2	BS		C6
Spotted Frog (east of Cascade Range)	Pr-S/u	G4/S2			
Northern Leopard Frog	Pr-S/c	G5/S2?	AS		
Foothill Yellow-legged Frog	Pr-S/v	G3/S3?	TS		
Bullfrog (introduced)	gamefish	G5/SE			

STATUS OF AMPHIBIANS IN WASHINGTON: JUNE 1996

SPECIES	WDFW	BLM	USFS	USFWS	TNC
Northwestern Salamander					G5/S5
Long-toed Salamander					G5/S5
Tiger Salamander	SM				G5/S4
Roughskin Newt					G5/S5
Cope's Giant Salamander	SM	AS	S		G3/S3
Pacific Giant Salamander					G5/S5
Olympic Torrent Salamander	SM	TS			G2G3/S2
Columbia Torrent Salamander	SC				G2/S2
Cascade Torrent Salamander	SC				G2/S2
Ensatina					G5/S5
Dunn's Salamander	SC				G4/S2
Larch Mountain Salamander	SS	ROD	ROD		G2/S2
Van Dyke's Salamander	SC	ROD	ROD		G3/S2
Western Redback Salamander					G5/S5
Tailed Frog	SM	AS			G3/S4
Great Basin Spadefoot					G5/S5
Western Toad					G4/S5?
Woodhouse's Toad	SM				G5/S3
Pacific Treefrog					G5/S5
Red-legged Frog		TS	S		G4/S5?
Cascades Frog		AS			G4/S4?
Spotted Frog	SC	BS		C6*	G4/S4*
Northern Leopard Frog	SC	AS			G5/S1
Bullfrog (introduced)	gamefish				G5/SE
Green Frog (introduced)					G5/SE

* west coast populations

WDFW: Washington Department of Fish & Wildlife (as of June 1996)

Species of special concern

SS = State sensitive
    (legally protected from taking)
SC = State candidate
SM = State monitor.

BLM: Bureau of Land Management, USDI

TS = Tracking species
AS = Assessment species
BS = Bureau sensitive
ROD = 'Survey & manage' species described in the Record of Decision for Amendments to USFS and BLM Planning Documents within Range of Northern Spotted Owl, April 1994.

USFS: Forest Service, Region 6, USDA

S = Sensitive
ROD = 'Survey & manage' species described in the Record of Decision for Amendments to USFS and BLM Planning Documents within Range of Northern Spotted Owl, April 1994.

USFWS: Fish & Wildlife Service, USDI (as of February 1996)

C = Candidate, sufficient information (number refers to priority).

TNC: The Nature Conservancy (as of December 1995)

G = Global (status throughout its range)
S = Subnational (status within Washington)
    1 = critically imperiled
    2 = imperiled
    3 = rare or uncommon
    4 = not rare, long-term concern
    5 = widespread, abundant, secure
    E = exotic
    ? = insufficient data.

STATUS OF AMPHIBIANS IN BRITISH COLUMBIA: JUNE 1996

All amphibian species native to B.C. are protected from taking, except by permit or regulation, under the *Wildlife Act* of 1982.

BCWB: B.C. Ministry of Environment, Lands & Parks, Wildlife Branch

R = Red List (species being considered for legal designation as Endangered or Threatened)

B = Blue List (vulnerable or sensitive)

Y = Yellow List (not at risk)

e = managed through ecosystem management

c = conservation species

g = global responsibility

i = non-native.

(Note: The Blue and Yellow list designations are currently under review.)

BC-CDC: B.C. Conservation Data Centre (equivalent to The Nature Conservancy's listings in Oregon and Washington)

G = Global (status throughout its range)

S = Subnational (status within B.C.)

1 = critically imperiled

2 = imperiled

3 = rare or uncommon

4 = frequent to common; restricted distribution; perceived future threats

5 = common, secure under present conditions

E = exotic.

COSEWIC: Committee on the Status of Endangered Wildlife in Canada.

v = vulnerable.

SPECIES	BCWB	BC-CDC	COSEWIC
Northwestern Salamander	Ycg	G5/S4S5	
Long-toed Salamander	Ye	G5/S5	
Tiger Salamander	R	G5/S2	
Roughskin Newt	Ye	G5/S4S5	
Pacific Giant Salamander	R	G5/S2	v
Ensatina	Yc	G5/S4	
Clouded Salamander	Yc	G5/S4	
Coeur d'Alene Salamander	R	G3/S1	
Western Redback Salamander	Yc	G5/S4	
Tailed Frog	B	G3/S3	
Great Basin Spadefoot	B	G5/S2S3	
Western Toad	Ycg	G4/S4	
Pacific Treefrog	Ye	G5/S5	
Striped Chorus Frog	Yc	G5/S4	
Red-legged Frog	Yc	G4/S4	
Spotted Frog (lower mainland)	R	G4/S1	
Spotted Frog (rest of B.C.)	Ycg	G4/S4	
Wood Frog	Ye	G5/S5	
Northern Leopard Frog	R	G5/S1	
Bullfrog (introduced)	i	G5/SE	
Green Frog (introduced)	i	G5/SE	

GLOSSARY

adhesive glands: glands that secrete a sticky fluid to enable an animal to cling to surfaces; two visible bumps under the chin of hatchling tadpoles that are resorbed within a few days of hatching.

adult: a sexually mature individual; usually the metamorphosed or terrestrial form, but it also includes the neotenic larvae of certain salamanders.

aquatic: having to do with water.

balancer: a whisker-like appendage from the lower part of each side of the head of some hatchling salamanders; provides a stable support before the front legs develop.

costal grooves: furrows on the sides of some salamanders that make the body appear segmented; occur between the intercostal folds.

dorsal: the upper surface (the top of the head, the back and the upper surface of the tail).

dorsal fin: a long, thin boneless fin that extends along the top of the tail and often the back of tadpoles and salamander larvae.

dorsolateral: both the back and the sides.

dorsolateral folds: thin, raised ridges that occur along the edges of the back of some frogs.

elk (or bear) wallows: water-filled depressions created by elk when they thrash about in bogs or wet meadows during the rutting season (or created by bears).

ephemeral: temporary, seasonal, lasting a very short time.

froglet: a juvenile frog that has just metamorphosed from a tadpole; it may still show the remains of a tail.

gill filaments: hair-like projections from the sides of each gill stalk; they make the gills appear feathery or comb-like.

gill stalk: the central shaft or prominent structure of a gill.

gills: structures (made up of stalks and filaments) that project from the back corners of the head of a larval salamander (or hatchling tadpole); they are used for respiration in water.

glands: see *poison glands*.

groin: the area where the thigh and belly meet on the side of a frog or toad; it is hidden when the hind leg is folded.

hatchling: the first larval stage of amphibians that have just hatched from eggs; this includes the early larval stage of pond-breeding salamanders (before the hind legs are visible), and the early tadpoles of frogs and toads (before the spiracle or eyes have developed, and while gills are still visible externally).

herpetology: the study of amphibians and reptiles, which are sometimes referred to as 'herptiles.'

intercostal folds: ridges on the sides of some salamanders that make the body appear segmented; occur between the costal grooves.

juvenile: froglets, toadlets and young salamanders that have metamorphosed or are in a terrestrial form, but are not yet sexually mature; juveniles have the same body form as terrestrial adults of their species, but they usually differ in proportions and color; includes the hatchlings of fully terrestrial salamanders.

larva: the free-swimming, aquatic form of an amphibian before metamorphosis; a salamander with gills or the tadpole of frogs and toads.

lateral: the sides.

lateral lines: lines of tiny pits along each side of the head and back of tadpoles that reflect a sensory system in the skin.

metamorphosis: the process of change from a larval (aquatic) stage of development to a form that can live on land.

nasolabial groove: a hair-thin furrow from the nostril down to the upper lip on some salamanders.

neotenic: sexually mature while retaining larval or immature characteristics (gills and fins); a neotene is a mature salamander that did not metamorphose.

nubbin: a small bump, point or knob.

parotoid areas: the back corners of the top of the head behind the eyes; the location of poison glands on some salamanders, frogs and toads.

perennial: permanent; retaining water year-round.

poison glands: glands under the skin that secrete toxins; visible as pores, pits or grainy areas.

riparian: occurring along the side of a stream, pond or lake; usually refers to vegetation that is influenced by the water level.

snout-to-vent length (SVL): the length of the head and body; the measurement from the end of the snout to the front corner of the vent.

spiracle: the tube-like opening for respiration on a tadpole; it is below the mouth on Tailed Frog tadpoles, and on the left side of all other frog and toad tadpoles.

tadpole: the larval, aquatic, early development stage of frogs and toads.

tail ridge: the upper surface of the tail trunk.

terrestrial: having to do with land.

toadlet: a juvenile toad that has just metamorphosed from a tadpole; it may still show the remains of a tail.

vent: the opening from the reproductive, intestinal and excretory systems; on most amphibians it is an inconspicuous slit, but it may be swollen, particularly on male salamanders during the breeding season.

ventral: the lower surface (the throat, the belly and the underside of the tail).

ventral fin: a long, thin, boneless fin that extends along the lower surface of the tail, behind the vent, on tadpoles and aquatic salamander larvae.

voucher: an accepted proof of occurrence kept at a museum; a set of photographs or a preserved specimen of a species that was observed or collected from a particular site.

COLORS AND MARKINGS

blotches: large, very irregularly shaped, may have indistinct edges.

blurred edge: indistinct or fading out (it looks like the edge on a drop of ink on a blotter).

dorsal stripe: a line on the back, the top of the head and the top of the tail of some amphibians that has a different color from the body.

dots: small, round, with well-defined edges.

flecks: tiny, angular or irregularly shaped, often metallic, with well-defined edges.

lip line: a light-colored stripe below the eye and above the mouth of many frogs.

markings: irregular in size and shape; not conforming to any of the other pattern types.

mask: a dark-colored patch behind the eye of some frogs; it may extend forward to the end of the snout and back to the shoulder or onto the side.

A dorsal stripe

metallic: shiny, looks like metal.

mottling: an interconnected network of long patches of one color on a background or matrix of another color.

network: dots or patches connected to each other by fine lines of the same color.

opaque: dense or colored so that no light comes through; can not be seen through.

A vague dorsal stripe composed of brown dots

patches: medium-sized to large, angular or irregularly shaped, with well-defined edges.

polka-dot: evenly and widely spaced spots.

scalloped edge: wavy, regularly indented, well defined.

spots: small to medium-sized, round, with well-defined or indistinct edges.

speckling: a dense, irregular arrangement of flecks or dots, often of different colors.

A pale yellow blotch and white flecks

tone: the general effect of the color; the overall tint or shade of color.

translucent: vaguely clear; light can be seen through.

transparent: colorless, clear; can be seen through.

wash: a color on top of another color.

A cream lip line

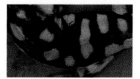

A sharply defined, black mask

Black polka-dots and metallic gold flecks

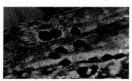

Crisp-edged black spots

A mottling of yellowish patches on a black backgound

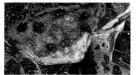

Spots with blurred edges

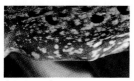

A green wash

REFERENCES

Only those publications that were used in the preparation of this manual, or that are most useful for surveying and identifying amphibians and their habitats in Oregon, Washington and British Columbia, are included.

Altig, R.G. 1970. A key to the tadpoles of the continental United States and Canada. *Herpetologica* 26 (2):180–207.

Altig, R.G., and P.H. Ireland. 1984. A key to salamander larvae and larviform adults of the United States and Canada. *Herpetologica* 40 (2):212–218.

Behler, J.L., and F.W. King. 1979. *The Audubon Society Field Guide to North American Reptiles and Amphibians.* Alfred A. Knopf, New York.

Bury, R.B., and P.S. Corn. 1991. *Sampling Methods for Amphibians in Streams in the Pacific Northwest.* General Technical Report PNW-GTR-275. U.S.D.A. Forest Service, Pacific Northwest Research Station, Portland.

Collins, J.T. 1990. *Standard Common and Current Scientific Names for North American Amphibians and Reptiles.* Herpetological Circular no. 19. Society for the Study of Amphibians and Reptiles, Lawrence, Kansas.

Corn, P.S., and R.B. Bury. 1990. *Sampling Methods for Terrestrial Amphibians and Reptiles.* General Technical Report PNW-GTR-256. U.S.D.A. Forest Service, Pacific Northwest Research Station, Portland.

Cowardin, L.M. 1979. *Classification of Wetlands and Deepwater Habitats of the United States.* FWS/OBS-79/31. U.S. Fish and Wildlife.

Davidson, C. 1995. *Frog and Toad Calls of the Pacific Coast.* Library of Natural Sounds, Cornell Laboratory of Ornithology, Ithaca, New York. Sound recording.

Green, D.M., T.F. Sharbel, J. Kearsley and H. Kaiser. 1997. Postglacial range fluctuation, genetic subdivision and speciation in the western North American spotted frog complex, *Rana pretiosa. Evolution.*

Green, D.M., and R.W. Campbell. 1984. *The Amphibians of British Columbia.* Royal British Columbia Museum Handbook no. 45. Victoria.

Hayes, M.P. 1994. *The Spotted Frog in Western Oregon.* Technical Report 94-1-01. Oregon Department of Fish and Wildlife, Wildlife Diversity Program, Portland.

Heyer, W.R., M.A. Donnelly, R.W. McDiarmid, L.C. Hayek and M.S. Foster. 1994. *Measuring and Monitoring Biological Diversity: Standard Methods for Amphibians.* Smithsonian Institution Press, Washington, D.C.

Jaeger, E.C. 1978. *A Source-book of Biological Names and Terms.* 3rd ed. Charles C. Thomas, publisher, Springfield.

Kirk, J.J. 1983. *Distribution of the Larch Mtn. Salamander* (Plethodon larselli) *in Oregon: With Notes on Other Plethodontids.* Nongame Technical Report 83-7-01. Oregon Department of Fish and Wildlife, Portland.

Leonard, W.P., H.A. Brown, L.L.C. Jones, K.R. McAllister and R.M. Storm. 1993. *Amphibians of Washington and Oregon.* Seattle Audubon Society, Seattle.

Marshall, D.B. 1992. *Sensitive Vertebrates of Oregon.* Oregon Department of Fish and Wildlife, Portland.

———. 1993. *Oregon Wildlife Diversity Plan, 1993-1998.* Oregon Department of Fish and Wildlife, Portland.

Matthews, D. 1988. *Cascade-Olympic Natural History.* Raven Editions, Portland.

McAllister, K.R. 1995. Distribution of amphibians and reptiles in Washington State. *Northwest Fauna* (Society for Northwestern Vertebrate Biology) 3:81–112.

Nussbaum, R.A., E.D. Brodie, Jr., and R.M. Storm. 1983. *Amphibians and Reptiles of the Pacific Northwest*. University Press of Idaho, Moscow.

Oregon Natural Heritage Data Base. 1989. *Rare, Threatened and Endangered Plants and Animals of Oregon*. Oregon Natural Heritage Data Base, Portland.

Ruggiero, L.F., K.B. Aubry, A.B. Carey and M.H. Huff (tech. coord.). 1991. *Wildlife and Vegetation of Unmanaged Douglas-fir Forests*. General Technical Report PNW-GTR-285. U.S.D.A. Forest Service, Pacific Northwest Research Station, Portland.

St. John, A.D. 1982a. *The Herpetology of Curry County, Oregon*. Nongame Technical Report 82-2-04. Oregon Department of Fish and Wildlife, Portland.

———. 1982b. *The Herpetology of Wenaha Wildlife Area*. Nongame Technical Report 82-4-03. Oregon Department of Fish and Wildlife, Portland.

———. 1984a. *The Herpetology of Jackson and Josephine Counties*. Nongame Technical Report 84-2-05. Oregon Department of Fish and Wildlife, Portland.

———. 1984b. *The Herpetology of the Upper John Day River Drainage*. Nongame Technical Report 84-4-05. Oregon Department of Fish and Wildlife, Portland.

———. 1985a. *The Herpetology of the Interior Umpqua River Drainage*. Nongame Technical Report 85-2-02. Oregon Department of Fish and Wildlife, Portland.

———. 1985b. *The Herpetology of the Owyhee River Drainage, Malheur County, Oregon*. Nongame Technical Report 85-5-03. Oregon Department of Fish and Wildlife, Portland.

———. 1986. *The Herpetology of the Willamette Valley*. Nongame Technical Report 86-1-02. Oregon Department of Fish and Wildlife, Portland.

———. 1987. *The Herpetology of Southwestern Klamath County*. Nongame Technical Report 87-3-01. Oregon Department of Fish and Wildlife, Portland.

Stebbins, R. C. 1985. *A Field Guide to Western Reptiles and Amphibians*. Houghton Mifflin Company, Boston.

Thoms, C., C.C. Corkran and D.H. Olson. 1997. Basic amphibian survey for inventory and monitoring in lentic habitats. *In* Sampling Amphibians in Lentic Habitats, eds. D.H. Olson, W.P. Leonard and R.B. Bury. *Northwest Fauna No. 4* (Society for Northwestern Vertebrate Biology). Olympia.

West, L., and W. Leonard. In press. *How to photograph Amphibians and Reptiles*. Stackpole Books.

Wildlife Branch and Habitat Protection Branch, B.C. Ministry of Environment, Lands and Parks. 1995. *Amphibians, Reptiles, Birds and Mammals not at Risk in British Columbia: The Yellow List (1994)*. Wildlife Bulletin no. B-74. British Columbia Ministry of Environment, Lands and Parks, Wildlife Branch, Victoria.

INDEX

COMMON NAMES

SCIENTIFIC NAMES

ABOUT THE AUTHORS

Char Corkran is an independent wildlife consultant. Her scientific curiosity was inspired by professors in biology, paleontology and literature at Brown University, where she received a Bachelor of Arts degree. Char's professional field work has focused on amphibians since 1990. Education and environmental concerns also compete for her time. She has written many reports and articles on wildlife subjects, and she has received several awards and grants for her wildlife conservation and volunteer work.

Chris Thoms is a certified Professional Wetland Scientist, through the Society of Wetland Scientists, a career sparked by her training in geomorphology and botany. Her Bachelor of Science degree from Portland State University focused on the natural sciences, and her work in amphibian ecology evolved along with her wetland specialty. Chris finds pleasure in the systematic observation of nature, and she has been a natural science illustrator and field naturalist since 1978.

Char and Chris discovered their parallel interests in 1980, and they worked together as field assistants on several projects through the Northwest Ecological Research Institute. They were unable to find adequate information on larval forms of amphibians, however, so together they designed field methods for identifying the early development stages of these engaging animals.

Chris Thoms and Char Corkran

Photograph by Jeffrey Kee